To Lowell with
warm best regards.
Grangor

CARBON CAPTURE AND SEQUESTRATION

CARBON CAPTURE AND SEQUESTRATION

Removing the Legal and Regulatory Barriers

M. Granger Morgan, Sean T. McCoy
and
Jay Apt, Michael Dworkin, Paul S. Fischbeck, David Gerard, Kaitlin A. Gregg, R. Lee Gresham, Colin R. Hagan, Donald M. Kreis, Robert R. Nordhaus, Emily R. Pitlick, Melisa Pollak, Jessica L. Reiss, Edward S. Rubin, Kari Twaite, and Elizabeth J. Wilson

RFF PRESS
RESOURCES FOR THE FUTURE

First published 2012
by RFF Press
Routledge, 711 Third Avenue, New York, NY 10017

Simultaneously published in the UK and ROW
by RFF Press
Routledge, 2 Park Square, Milton Park, Abingdon, Oxon OX14 4RN

RFF Press is an imprint of the Taylor & Francis Group, *an informa business*

The findings, interpretations, and conclusions offered in RFF Press publications are those of the authors. They do not necessarily represent the views of Resources for the Future, its directors, or its officers. Similarly, any geographic boundaries and titles depicted in RFF Press publications do not imply any judgment or opinion about the legal status of a territory on the part of Resources for the Future.

Library of Congress Cataloging in Publication Data
Carbon capture and sequestration: removing the legal and regulatory barriers / by M. Granger Morgan . . . [et al.].
p. cm.
1. Carbon sequestration—Law and legislation—United States. I. Morgan, M. Granger (Millett Granger), 1941–
KF3812.2.C37 2012
344.7304'6342—dc23 2011049561

ISBN: 978-1-61726-101-5 (hbk)
ISBN: 978-0-203-11505-3 (ebk)

Typeset in Bembo
by RefineCatch Limited, Bungay, Suffolk

Printed and bound in the United States of America by Sheridan Books, Inc.(a Sheridan Group company)

ABOUT RESOURCES FOR THE FUTURE *AND* RFF PRESS

Resources for the Future (RFF) improves environmental and natural resource policymaking worldwide through independent social science research of the highest caliber. Founded in 1952, RFF pioneered the application of economics as a tool for developing more effective policy about the use and conservation of natural resources. Its scholars continue to employ social science methods to analyze critical issues concerning pollution control, energy policy, land and water use, hazardous waste, climate change, biodiversity, and the environmental challenges of developing countries.

RFF Press supports the mission of RFF by publishing book-length works that present a broad range of approaches to the study of natural resources and the environment. Its authors and editors include RFF staff, researchers from the larger academic and policy communities, and journalists. Audiences for publications by RFF Press include all of the participants in the policymaking process—scholars, the media, advocacy groups, NGOs, professionals in business and government, and the public.

Disclaimer

Opinions expressed in this book are those of the authors and may not represent the view of the institutions with which they are affiliated.

CONTENTS

FIGURES AND TABLES

Figures

Tables

AUTHORS

Jay Apt is a Professor in Carnegie Mellon University's Tepper School of Business and in the Carnegie Mellon Department of Engineering and Public Policy. He directs the Carnegie Mellon Electricity Industry Center. He has received the NASA Distinguished Service Medal and the Metcalf Lifetime Achievement Award for significant contributions to engineering. He holds an AB in physics from Harvard College and a PhD in experimental atomic physics from the Massachusetts Institute of Technology.

Michael Dworkin is Professor of Law and Director of the Institute for Energy and the Environment at the Vermont Law School. He clerked for the DC Court of Appeals, represented US EPA in appellate litigation, and was general counsel and then chairman of the Vermont Public Service Board. He is a director of the board of the Vermont Energy Investment Corporation and served on the boards of the EPRI and the American Council for an Energy Efficient Economy. He earned a BA from Middlebury College and a JD from Harvard Law School.

Paul S. Fischbeck is a Professor in the Department of Engineering and Public Policy and the Department of Social and Decision Sciences at Carnegie Mellon University. His research involves normative and descriptive risk analysis on topics ranging from offshore oil production platforms, to air pollution from international shipping, and the space shuttle's tile protection system. He holds degrees from the University of Virginia (BS), the Naval Postgraduate School (MS), and Stanford (PhD).

David Gerard is an Associate Professor of Economics at Lawrence University. He was previously Executive Director of the Center for the Study and Improvement of Regulation in the Department of Engineering and Public Policy at Carnegie Mellon University. His area of expertise is in risk regulation and focuses on the interrelationships between regulation and technological change. He earned a BA from Grinnell College and an MS and a PhD in economics from the University of Illinois.

Kaitlin A. Gregg is an attorney at Van Ness Feldman, PC, a law firm specializing in environmental and energy law based in Washington, DC. Prior to joining Van Ness Feldman, Kaitlin was a law clerk for the Honorable William K. Sessions, a federal district judge in Vermont. Kaitlin holds a law degree from Yale Law School, a bachelor's degree from Middlebury College, and a master's degree from the Yale School of Forestry and Environmental Studies.

R. Lee Gresham is a Natural Gas and Power Markets Consultant at SAIC Energy, Environment, and Infrastructure where he works in the Economic Modeling and Asset Transaction practice groups. His work focuses on the interplay between electric generation technologies and fuels, economics, regulation, and policy. Lee holds a PhD in Engineering and Public Policy from Carnegie Mellon University, a JD from Vermont Law School, and a BA in Psychology from the College of the Holy Cross.

Colin R. Hagan is a law student and Research Associate at the Institute for Energy and the Environment at Vermont Law School. A graduate of Furman University, Colin has a background in federal climate change and energy policy. He recently served as a summer law clerk at the Clean Air Task Force and a Summer Honors Clerk at the US Environmental Protection Agency's Office of General Counsel. Colin is a 2011 Switzer Environmental Fellow.

Don M. Kreis is Professor of Law and Associate Director of the Institute for Energy and the Environment at Vermont Law School. He served as general council to the New Hampshire Utilities Commission and in judicial clerkships with the Vermont Superior Court, the US District Court for the District of Maine, and the Maine Supreme Judicial Court. He holds a BA from Middlebury, a JD from the University of Maine Law School, and an MS from Columbia University's School of Journalism.

Sean T. McCoy is an Adjunct Assistant Professor at the Department of Engineering and Public Policy at Carnegie Mellon University and an Energy Analyst at the International Energy Agency in Paris, France. The focus of Sean's research is the interaction between regulation, energy technologies, and energy resources. Sean holds a PhD in Engineering and Public Policy from Carnegie Mellon University and a BASc from the University of Waterloo in Environmental Engineering (Chemical Specialization).

M. Granger Morgan is Professor and Head of the Department of Engineering and Public Policy at Carnegie Mellon University. His research addresses problems in science, technology, and public policy in which the technical details and uncertainty are of central importance. He is a member of the U.S. National Academy of Sciences and a Fellow of the AAAS, the IEEE, and the SRA. He holds degrees from Harvard (BS), Cornell (MS), and UCSD (PhD, applied physics).

Robert R. Nordhaus is a member of the law firm of Van Ness Feldman, PC, and an Adjunct Professor at the George Washington University Law School. He was a member of the Energy Policy and Planning Office in the Carter White House and was counsel to the House Interstate and Foreign Commerce Committee. After three years as the Federal Energy Regulatory Commission's first General Counsel, he joined Van Ness Feldman in 1981 and practiced there until 1993, when he was appointed General Counsel of the Department of Energy by President Clinton. He rejoined the firm in 1997.

Emily R. Pitlick is an attorney at Van Ness Feldman, PC, a law firm specializing in environmental and energy law based in Washington, DC. At Van Ness Feldman, she provides counsel to gas and oil pipeline companies on their federal regulatory obligations. She holds a law degree from Duke University School of Law, and a bachelor's degree from Washington University in St. Louis.

Melisa Pollak is a Research Fellow at the Humphrey School of Public Affairs at the University of Minnesota. She studies low carbon energy systems, focusing on the interplay between technical and sociopolitical factors. Melisa holds an MS in Science, Technology, and Environmental Policy and a BS in Geophysics from the University of Minnesota.

Jessica L. Reiss works as a senior environmental consultant, assisting federal agencies with sustainability, environmental compliance, and environmental management. Prior to law school, she worked as an Environmental Health and Safety Coordinator throughout the Midwest. Jessica is licensed to practice law in the State of Indiana. She earned a BS from Ball State University and a Juris Doctor, Master of Environmental Law and Policy, and a Certificate in Energy Law from Vermont Law School.

Edward S. Rubin is the Alumni Chair Professor of Environmental Engineering and Science and a professor in the Departments of Engineering and Public Policy, and Mechanical Engineering at Carnegie Mellon University. His research is in the areas of energy utilization, environmental control, technology innovation, and technology-policy interactions. He earned a BS in mechanical engineering from the City College of New York and MS and PhD degrees from Stanford University.

Kari Twaite is an Associate in the Washington, DC office of Beveridge & Diamond, PC. She earned a BA from Bowdoin College in 2006, a JD from Vermont Law School, and a Master of Environmental Management from the Yale University School of Forestry and Environmental Studies in 2011. She interned with the US Department of Energy's Office of General Counsel, with Vermont Institute for Energy and Environment, and the Yale Center for Environmental Law and Policy.

Elizabeth J. Wilson is an Associate Professor of Energy and Environmental Policy and Law at the Humphrey School of Public Affairs at the University of Minnesota. Her research examines policies and institutions supporting development of carbon-managed energy systems. She holds a doctorate in Engineering and Public Policy from Carnegie Mellon University and was selected as a Leopold Leadership Fellow in 2011.

PREFACE

For many years, investigators in the Department of Engineering and Public Policy at Carnegie Mellon University have done engineering and economic analysis of technologies to capture carbon dioxide and sequester it permanently in appropriate deep geological reservoirs. In recent years, it became increasingly apparent that, while technical improvements and commercial-scale demonstration are important, carbon capture and geologic sequestration (CCS) will not be able to make a significant contribution to reducing US emissions of carbon dioxide unless profound changes occur in the regulatory environment.

In January 2008, this realization led to the creation of the CCS Regulatory (CCSReg) Project, an interdisciplinary project to design a regulatory environment to support the capture, transport, and geologic sequestration of carbon dioxide in a manner that is safe, environmentally sound, affordable, compatible with evolving international carbon control regimes (including emissions trading), and socially equitable.

Anchored at Carnegie Mellon University, CCSReg has involved an extended collaboration between technical and policy experts in the Department of Engineering and Public Policy at Carnegie Mellon University (www.epp.cmu.edu) and in the Humphrey School of Public Affairs at the University of Minnesota (www.hhh.umn.edu), and with legal experts at the Vermont Law School (www.vermontlaw.edu) and the Washington, DC law firm of Van Ness Feldman (www.vnf.com).

In addition to being both enjoyable and educational for all of those involved, this collaboration has been remarkably productive, and has led to a set of insights and recommendations for how the US should undertake the regulation of CCS. This book summarizes what we have learned and the Appendix provides a concrete illustration of how our recommendations might be converted into a new legal and regulatory framework for the nation.

Once the US becomes serious about achieving a major reduction in its emissions of carbon dioxide, we see no cost-effective way that this can happen without CCS playing a significant role for at least the first several decades. It is our deep hope that the ideas developed in this book will help to make that possible.

ACKNOWLEDGMENTS

While the authors are solely responsible for the content of this book, we received help, critical advice, and support from a number of people. We thank Devorah Ancel, Allyson Anderson, Owen Anderson, Barbara Bugosh, Katelyn Bush, Melisa Chan, Denise Caruso, Thomas Feeley, Lauren Fleishman, Sharon Grant, Edward Helminski, Emily Fisher, Julio Friedman, Sharon Grant, Michael Griffin, Colin Hagen, Patrick Joy, Mark Latham, Lester Lave, Aaron Lotlikar, Jennifer Miller, Karen Obenshain, Lindene Patton, Koring Sharp, Patti Steranchak, Kari Twaite, Philip Marston, Paul Parfomak, John Tombari, Chiara Trabucchi, Emily Whitmore, and Ben Yamagata.

Primary financial support for this work was provided by a grant from the Doris Duke Charitable Foundation (Grant 2007117) to the Department of Engineering and Public Policy at Carnegie Mellon University. Additional support was provided by the Climate Decision Making Center (SES-0345798) and by the Center for Climate and Energy Decision Making (SES-0949710), both through a cooperative agreement between the National Science Foundation and Carnegie Mellon University; from Carnegie Mellon's Electricity Industry Center; and from the Department of Engineering and Public Policy at Carnegie Mellon University.

ABBREVIATIONS

AoR	Area of review of an injection site (under UIC rules)
BLM	Bureau of Land Management
CCS	Carbon capture with geologic sequestration
CO_2	Carbon dioxide
CTL	Coal to liquids
DOE	The US Department of Energy
DOT	The Department of Transportation
EIA	The Energy Information Agency
EOR	Enhanced oil recovery
EPA	The US Environmental Protection Agency
FERC	The Federal Energy Regulatory Commission
FGSB	The new Federal Geologic Sequestration Board proposed in this book
GAO	The General Accounting (now Accountability) Office
GS	Geologic sequestration
GSTAC	Geologic Sequestration Technical Advisory Committee of the NRC that is proposed in this book
GHG	Greenhouse gas
HLPSA	Hazardous Liquid Pipeline Safety Act
ICA	Interstate Commerce Act
IGCC	Integrated Gasification Combined Cycle
IPCC	Intergovernmental Panel on Climate Change
IOGCC	Interstate Oil and Gas Compact Commission
LCOE	Levelized cost of electricity
MLA	Mineral Leasing Act
MRV plan	Monitoring, reporting and verification plan
NAAQS	National Ambient Air Quality Standard

NGA	Natural Gas Act
NRC	The US National Research Council
PHMSA	The Pipeline Hazardous Materials Safety Administration
SDWA	US Safe Drinking Water Act
STB	The Surface Transportation Board
UIC	The US Underground Injection
USDW	Underground Source of Drinking Water

1

THE IMPORTANCE OF CARBON CAPTURE AND GEOLOGIC SEQUESTRATION IN A CARBON CONSTRAINED WORLD

The world's economy runs on energy. Today most of that energy is obtained by burning "fossil fuels," comprised mainly of coal, oil, and natural gas. Burning these fuels releases prodigious amounts of heat-trapping carbon dioxide (CO_2) into the atmosphere. Unless the world reduces its emissions of CO_2 by about 80% by the middle of this century, the future looks grim.[1] Our grandchildren may see the complete disappearance of summer sea ice in the arctic and the end of polar bears in the wild. As Figure 1.1 suggests, they may also witness the loss of sugar maples and most of the ski industry in New England.[2] Because sooner or later most CO_2 ends up in the ocean, by late in the century they can expect to see the end of most coral reefs[3] as shown in Figure 1.2, as well as the demise of many zooplankton at the bottom of food chains, which are a food source for salmon, whales, and other sea life.[4] If the worst happens, their grandchildren may see all of South Florida and the Gulf Coast disappear under rising sea levels, as shown in Figure 1.3.

In order to reduce atmospheric greenhouse gas concentrations by the amount necessary to avoid the worst consequences of climate change, industrialized and developing nations must reduce dramatically the emissions that result from burning coal, oil, and natural gas. This will require changing the way energy is produced and used around the world.

Box 1.1 *The Link between CO_2 Emissions and Concentration*

Why do we say that the world must reduce its emissions of CO_2 by roughly 80%? Carbon dioxide and the other greenhouse gases that produce global warming are not like conventional air pollutants such as sulfur dioxide (SO_2), oxides of nitrogen (NO_x), smog, and fine particles (PM). These conventional air pollutants only remain in the atmosphere for a few hours or days. As a result, the concentration of these pollutants decreases relatively rapidly once their emissions are decreased and, thus, their effects on the atmosphere are short-lived.

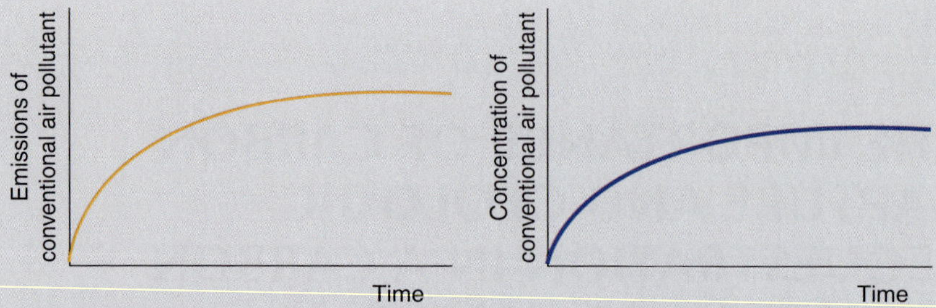

In contrast, most of the CO_2 and other greenhouse gases produced by human activities that enter the atmosphere remain there for a century or more. Even if we could halt the increase in greenhouse gas emissions, the atmospheric concentration of greenhouse gases that cause warming will continue to rise.

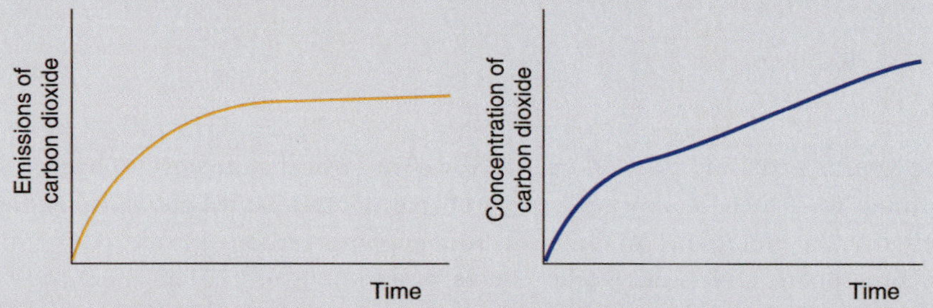

If we want to reduce atmospheric concentrations of CO_2 and other greenhouse gases, we must reduce their emissions by approximately 80%.

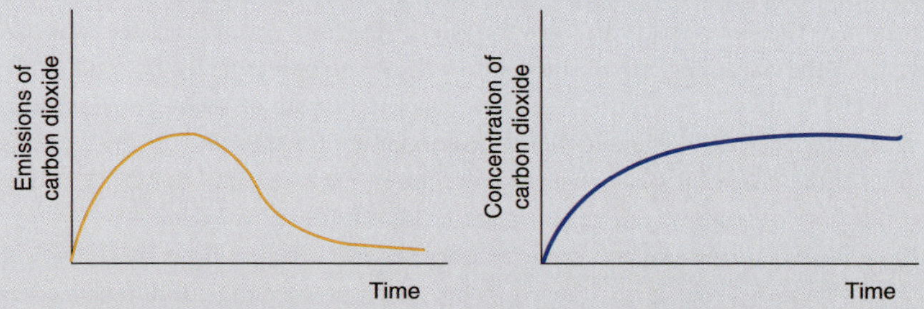

A useful analogy is a bathtub (the atmosphere) with a very large faucet (human emissions) and a much smaller drain (natural processes that remove CO_2 and other greenhouse gases from the atmosphere). Without us turning down the faucet significantly, the bathtub continues to fill up.

The rise of emissions in countries such as China and India makes turning down the faucet even more difficult. China's annual emissions of greenhouse gases recently passed those of the US, and emissions from India are also rapidly growing. However, it will be decades before these additions to the "global bathtub" of atmospheric greenhouse gases surpass the quantities that the US, Europe, Japan, and other developed countries have already poured in. Disagreements over the necessary quantity of emissions reductions depends on what people believe is a "safe level" of CO_2 in the atmosphere. The exact quantity (e.g., 80% versus 70% in 2050) is far less important.

FIGURE 1.1 *Graphical illustration of how the climates of New Hampshire and of upstate New York are likely to change based on an average of the GFDL, HadCM3, and PCM model projections. (Figure courtesy of NECIA.)*

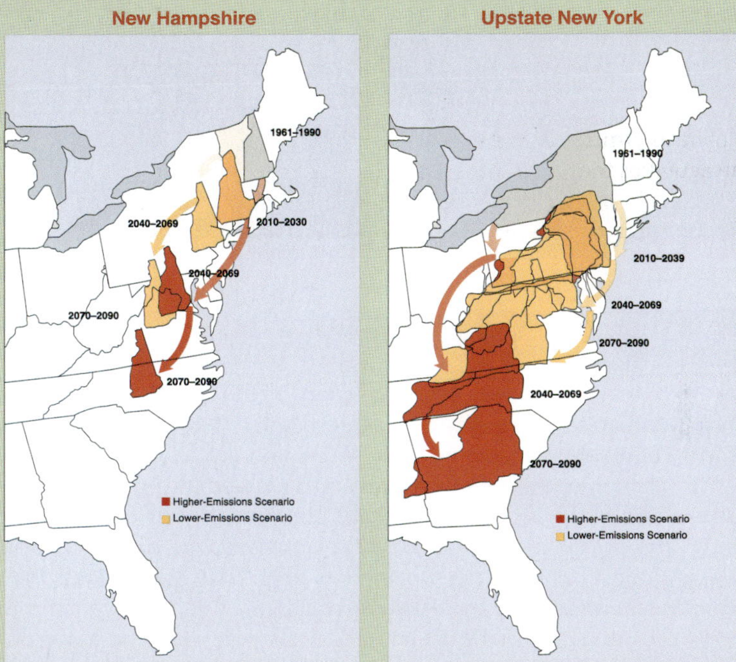

FIGURE 1.2 *Illustration of the likely impact on coral reefs from the rise in atmospheric concentrations of CO_2, and associated ocean acidification, that can be expected by later this century. (Figure courtesy of Hoegh-Guldberg et al.,* Science.*)*

FIGURE 1.3 *Illustration of the impact on the US Gulf Coast of losing half the ice covering Greenland, which would result in approximately 3.5 meters of sea-level rise. (Figure courtesy of Daniel Schrag, Harvard.)*

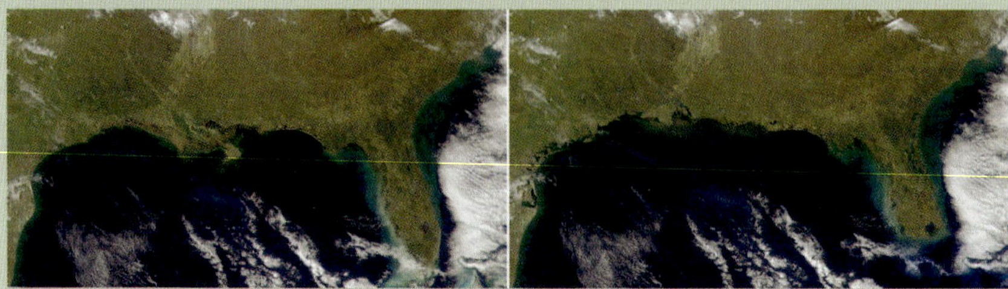

BOX 1.2 *Some Climate Change Basics*

Energy from the sun powers the circulation of the earth's atmosphere and oceans in addition to sustaining most life on earth. That energy largely arrives in the form of visible light. About 30% of it is immediately reflected back into space. This fraction of the light that is reflected is called the "planetary albedo." The other 70% of the incoming energy is absorbed by clouds, land, and the oceans. If that energy just continued to be absorbed, the earth would rapidly heat up, and most living species would be unable to survive. To keep the earth at a constant temperature, the 70% of solar input that is absorbed must be radiated back into space as heat (infrared). However, while the atmosphere is transparent to visible light, it is largely opaque to infrared. This is because water vapor, CO_2, and other trace gases that occur naturally in the atmosphere absorb infrared radiation, trapping the heat energy. As a consequence, the earth heats up until it reaches a temperature at which enough heat energy is radiated back into space from the top of the atmosphere to bring the outward flow of energy into balance with what is being absorbed. As a result of this "greenhouse warming," the average temperature of the earth is about 33°C (approximately 60°F) warmer than it would be if there was no atmosphere.

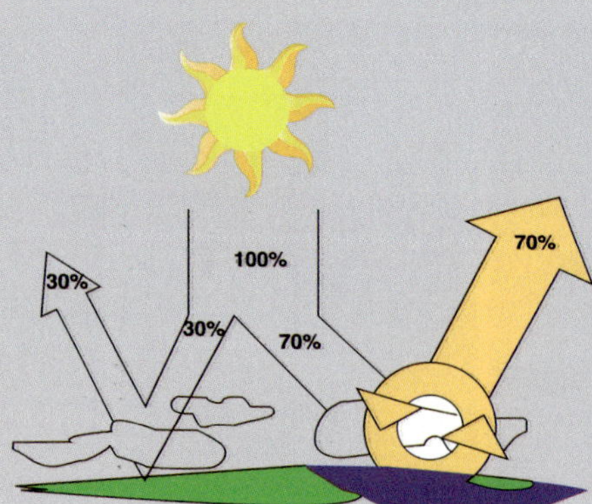

Human activity has been adding more CO_2 (and other greenhouse gases) to the atmosphere, which increases the naturally occurring greenhouse effect, causing warming and other changes in climate. Because CO_2, once released into the atmosphere, remains there for over 100 years, the atmospheric concentration of CO_2 has been steadily rising, and with it the average temperature of the earth. This, in turn, is leading to a variety of other changes in climate.

Details on climate change and its impact can be found in the reports of the Intergovernmental Panel on Climate Change at www.ipcc.ch. The National Oceanic and Atmospheric Administration provides additional information at www.noaa.gov/climate.html.

1.1 WHY DOES THE WORLD STILL NEED FOSSIL FUEL?

Many believe that when the US and the rest of the world do finally embark on major reductions in the emissions of CO_2, this could be achieved by simply switching to more energy efficient cars and appliances and converting to solar, wind, biomass, and nuclear power. All these can help, but each also has its practical and economic limits.

There are big opportunities to save energy through improved efficiency, often with savings or modest cost and with few drawbacks.[5] For example, lighting comprises 20% of all US electricity consumption. Use of compact fluorescents can significantly reduce energy consumption, and solid-state lighting (e.g., light emitting diodes or LEDs) offers the potential to further reduce energy consumption from lighting at lower cost.[6] With better design, buildings can be made *much* more energy efficient.[7] In the transportation sector, which is responsible for roughly one-third of greenhouse gas emissions in the US, regular hybrid and plug-in electric hybrid automobiles also show promise to reduce both emissions and our dependency on imported oil.[8]

While using energy more efficiently can contribute greatly to reducing emissions, energy must still come from something. Today, the US makes roughly half of all its electricity from coal and other fossil fuels. This fraction is even larger in other countries, as shown in Figure 1.4. Displacing fossil energy with renewable and nuclear energy sources will help, but each alternative has its practical and economic limits.

The big limitation on energy from wind and solar is that it is inherently variable and intermittent. The wind does not blow all the time. Even when many wind farms are connected together, intermittency is still a problem.[9] As it becomes cheaper, storage will help, but in the meantime, many utilities must use gas turbines to fill in the gaps when the wind fluctuates. This can cause large increases in emission of nitrogen oxide air pollutants.[10] Sunlight is only available during the day, and passing clouds can create considerable intermittency. Even in sunny Arizona, the output from solar arrays is a little less than 20% of what it would be under full sun 24 hours a day.[11] Although the cost of solar power continues to decline, it is still extremely

FIGURE 1.4 *Sources of generated electric power in the US and around the world in 2004. Numbers in parentheses are generation in billions of kilowatt-hours. Note the high dependency on coal and other fossil fuels. (Data from the US Energy Information Agency.)*

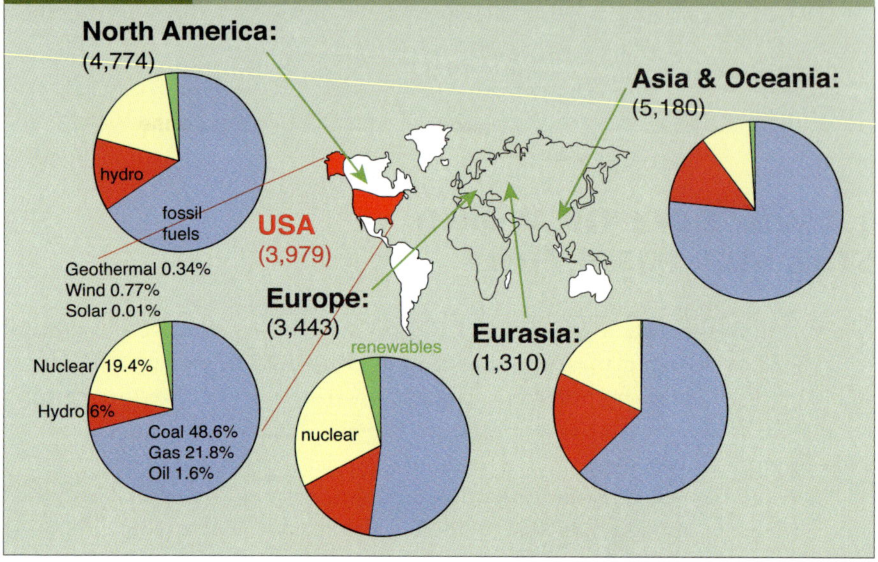

expensive and, barring unexpected technical breakthroughs, it appears likely to remain too expensive for bulk electricity supply for many decades.[12]

Biomass is another possible energy source for both electricity generation and transportation. However, biomass is not without drawbacks: the amount of land required to grow energy crops to supply a significant fraction of human energy requirements would have significant impacts on availability of land for food crops, and is already having impacts on world food prices.[13] In addition, the scientific understanding of the greenhouse gas implications of using biomass to generate electricity or transportation fuel continues to develop, but it has become clear that all biomass energy is not created equally.[14]

Nuclear power, which today produces about 20% of electricity consumed in the US, and a larger fraction in Europe, is expensive and faces other environmental and safety concerns.[15]

Existing investments in many conventional energy facilities are still being paid off and cannot be abandoned without creating large "stranded costs." Similarly, while many appliances have become more efficient, consumers often cannot afford to throw out their old ones and replace them overnight with more efficient versions. Even with public policies that could ease the financial burden and allow an accelerated pace of conversion, replacing all this "capital stock" will take time.

FIGURE 1.5 *Example of a cost-effective portfolio to achieve a 50% reduction in CO_2 emissions from the US electric power system by 2050. EE refers to improved energy efficiency. (Figure courtesy of EPRI.)*

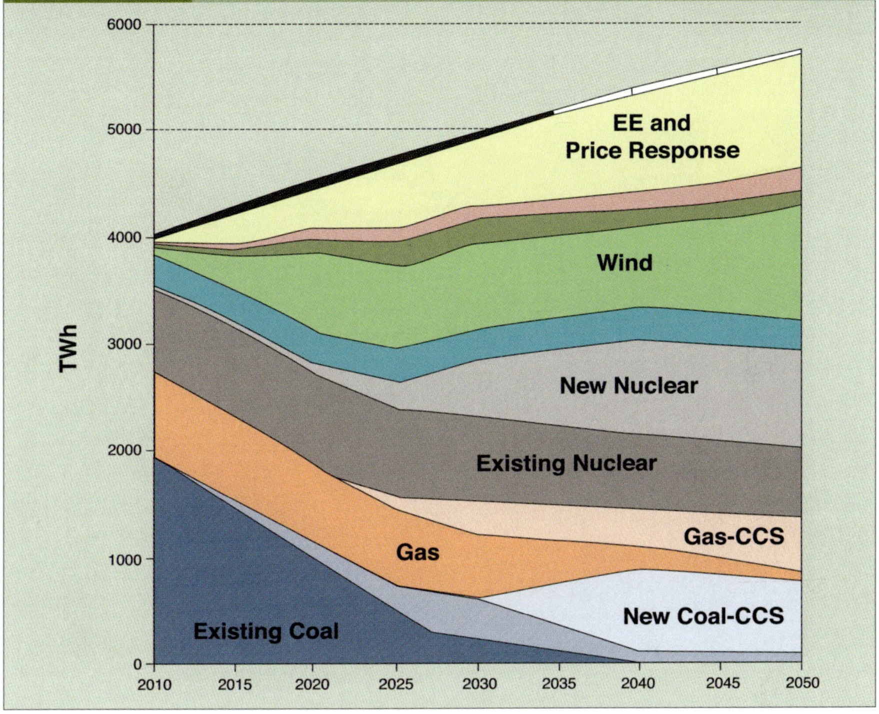

For all these reasons, there is no single "silver bullet" to decarbonize the world energy system.[16] Reducing dramatically CO_2 emissions in as affordable way as possible will require a *portfolio* of everything we have got on the supply side along with improved end-use efficiency. Figure 1.5 shows one example of what such a portfolio might look like. For at least the coming half century, there is simply no cost-effective way to avoid using fossil fuel as a part of a portfolio of energy solutions. This means that the US must develop ways to use fossil fuels, especially coal and natural gas for which the US has ample domestic supplies,[17] without releasing CO_2 to the atmosphere.

1.2 CARBON CAPTURE WITH GEOLOGIC SEQUESTRATION (CCS)

To get some idea of how much CO_2 we produce from electricity generation alone, consider just the one big power plant in Figure 1.6. Every day a plant this size

FIGURE 1.6 *Every day a coal-fired power plant of this size (2460 Mw) burns roughly as much coal as is carried in 200 railroad hopper cars (100 tons each) into carbon dioxide gas and releases it into the atmosphere. There are thousands of large coal plants like this all around the world.*

consumes all the coal that would be carried by 200 rail hopper cars (each of which carries 100 tons). All the carbon in that coal is converted into invisible CO_2 gas and released into the atmosphere. There are thousands of plants like this all around the world.

Fortunately, a handful of technologies exist today that can capture CO_2 from the combustion of fossil fuels such as coal. While most of these technologies exist today at commercial scale in other applications, they are only just now being applied to the capture of CO_2 from power plants and others sources.

In facilities such as power plants that have CCS,[18] captured CO_2 is compressed into a dense liquid (technically a supercritical fluid), and injected into appropriate geological formations a kilometer or more below the surface. There, the CO_2 is safely and permanently disposed, or "sequestered," from the atmosphere. CCS closes the carbon cycle since the carbon contained in the fossil fuels came from the ground, and is returned to the ground.

While CO_2 emissions-reduction is a global concern, this book focuses on how to achieve safe and permanent underground sequestration of CO_2 in the US. If the US intends to use CCS to make reductions in greenhouse gas emissions of billions of tons per year in the coming decades, regulations must ensure that CCS, and geologic sequestration in particular, is safe and permanent. The chapters that follow outline a regulatory framework designed to assure that CCS is done in a way that is safe, environmentally sound, affordable, socially equitable, and compatible with evolving international carbon control regimes (including emissions trading).

1.3 UNDERGROUND INJECTION TODAY

Today the US injects large quantities of waste fluid underground. Anyone who wishes to make such injections must obtain a permit under the EPA Underground Injection Control (UIC) program. As explained in Chapter 4, the EPA has delegated the authority to issue and oversee permits to many states. The rules governing injection are quite different for waste produced by different processes and injected under different "well classes." For example, waste fluids from the oil and gas industry (Class II wells) are treated much more leniently than the same fluid produced by other industrial processes (e.g., Class I wells). The UIC rules apply only to the operation of the well itself and the injection activity. They do not require any approval from the holders of surface or mineral rights. For example, the State of Florida injects roughly three billion tonnes per year of treated municipal wastewater with no formal approval from any property owners.

The US EPA has recently developed and promulgated a new well class (Class VI) to cover CCS. As with other well classes under the UIC program, these new regulations are silent on such key issues that arise in the CCS context—such as access to pore space, liability, and long-term stewardship. These issues are discussed at greater length in Chapter 4.

1.4 THE BOUNDARIES AND LIFE CYCLE OF A CCS PROJECT

The CCS process involves three steps: CO_2 capture, transport of captured CO_2 to the sequestration site, and geologic sequestration of the CO_2 in deep geologic formations (the last stage being abbreviated as GS in this book). These three steps are illustrated in Figure 1.7.

In the first step, CO_2 is captured from large point sources using any of a number of different processes. Once the gas has been captured it is then dried, compressed, and cooled to convert the gaseous CO_2 into a supercritical fluid. In step two, supercritical CO_2 is moved from the site of capture to the site of injection, typically by pipeline, although ocean-going tankers (similar to LNG tankers) could also be used where appropriate.[19] Once at the site of injection, CO_2 is injected into a geologic formation deep enough below the surface that it will remain as a supercritical fluid. This is the third and final step. The minimum depth which meets this criterion is typically 800 m (2,600 ft), although this varies from place to place.[20]

FIGURE 1.7 *The steps in the capture and deep geologic sequestration of CO_2.*

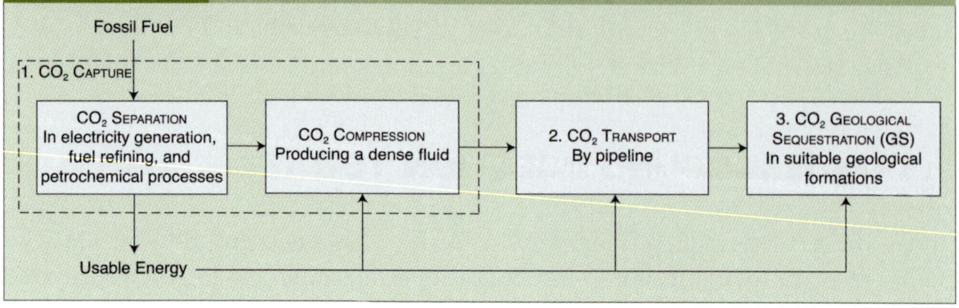

Both a power plant and the associated storage site (or sites) will require several years of lead time before full-scale operation can commence. However, in contrast to a power plant which can be removed at the end of its useful life and the land reclaimed, a sequestration site will require a long-term commitment after injection at the site ceases, during which time the site will need to be monitored to ensure that the injected CO_2 is behaving as expected.

The life cycle of a GS project is illustrated in Figure 1.8. We will refer to this diagram from time to time in later chapters of this book as we address issues that arise at various stages along a project's life cycle.

FIGURE 1.8 *Life cycle of a geologic sequestration project, illustrating not only the processes that occur during operation but also the overall stages and their approximate duration.*

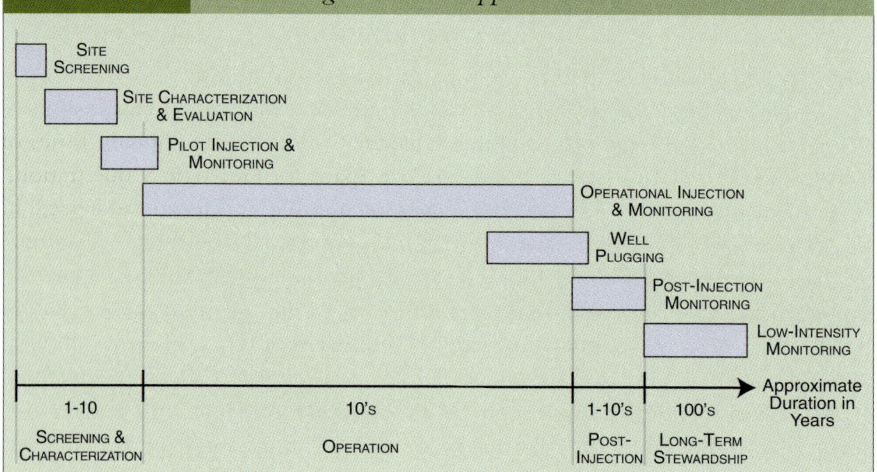

All of the phases of a CCS project outlined in Figure 1.8 will require regulatory oversight of some type to ensure that CCS is carried out safely and effectively. As discussed in greater detail in Chapter 4, the US Environmental Protection Agency (US EPA) oversees an Underground Injection Control Program (UIC) as part of its responsibilities under the US Safe Drinking Water Act (SDWA) (42 U.S.C. 300h et seq.); portions of the operational and post-injection phases are shown in this figure. However, these regulations are limited in scope by the SDWA and are silent on key issues that will impact CCS development, as discussed in later chapters.

1.5 THE REASON FOR THIS BOOK

The US likely cannot achieve deep reduction in emissions of CO_2 over the next fifty years without the wide adoption and use of CCS on coal- and gas-fired power plants and other industrial processes: this as part of a portfolio of strategies that also involve renewables, nuclear, and improved end-use efficiency. While additional technological advances will be required to achieve this goal, it will also require dramatic improvement and streamlining of what is now an incoherent and fractionated regulatory system.

The chapters that follow describe the set of regulatory issues that must be addressed to make the safe deployment of large scale CCS economically feasible and outline how each of these issues can best be resolved. Chapter 2, "Technology for Carbon Capture and Geologic Sequestration," reviews the technology required for carbon capture and geologic sequestration and, in particular, expands on the steps in the life cycle illustrated in Figure 1.8. The location of power plants is often not close to appropriate sites for geologic sequestration, so often CO_2 pipelines will be required. Chapter 3 addresses "Siting CO_2 Pipelines for Geologic Sequestration." Chapter 4, "Permitting Geologic Sequestration Sites," describes the current situation for regulation of CCS and identifies a set of regulatory gaps that must be addressed in order to create a coherent national framework for geologic sequestration. Then, in Chapter 5, "Learning from and Adapting to Changes in Geologic Sequestration Technology," we lay out the adaptive framework that we recommend for geologic sequestration. Chapter 6 addresses the difficult issues of assuring the projects can gain "Access to Pore Space for Geologic Sequestration." Issues of liability across the various stages of a project's life are addressed in Chapter 7, "Liability and the Management of Long-term Stewardship." Chapter 8, "Greenhouse Gas Accounting for CCS," reviews the issues that must be addressed to track the amount of CO_2 that is captured, transported, and sequestered. Chapter 9 explores some of the commercial and other considerations involved in "Making CCS a Reality." Finally, Chapter 10 briefly recaps the arguments presented throughout the book and summarizes our recommendations.

2

TECHNOLOGY FOR CARBON CAPTURE AND GEOLOGIC SEQUESTRATION

As the previous chapter briefly explained, the process of carbon capture and sequestration (CCS) includes three steps: capture of carbon dioxide (CO_2), transport of the captured CO_2 to an injection site, and injection of the CO_2 deep underground, where it is sequestered from the atmosphere. These steps were illustrated in Figure 1.7. In the first step, CO_2 is captured from large point sources using a capture technology suited to the facility (described in Sections 2.2, 2.3, and 2.4). This CO_2, together with small quantities of other gases derived from the combustion and capture processes, is then dried, compressed, and cooled. Depending on the temperature, it is converted from a gas to a liquid or supercritical fluid. This fluid is moved from the site of capture to the site of injection, typically by pipeline (described in Section 2.5), although tanker ships (similar to LNG tankers) could also be used in some cases.[1] Once it is delivered to the sequestration site, the CO_2 is injected into a geologic formation deep enough below the surface that the CO_2 remains as a supercritical fluid. On continents, assuming typical temperature and pressure gradients, the minimum depth that meets this criterion is typically 800 m or 2,600 ft. The precise minimum depth varies from one location to another.[2]

The facilities to which CO_2 capture will be applied, as well as infrastructure developed for transport and geologic sequestration (GS), are expensive and long-lived. For example, power plants that might install CO_2-capture systems typically have a life of at least thirty years, although in the US today many coal-fired power plants are much older (Figure 2.1). Carbon dioxide captured from such plants will be sent to one or more storage sites having an operational life of a similar span, through a pipeline network that will have an even longer useful life. However, in contrast to a power plant or pipeline, geology suitable for sequestration must be identified through exploration activities before a storage site is developed. Furthermore, development of a sequestration site requires a long-term commitment

Figure 2.1 *Age distribution of coal-fired power plants in the US. (Data from US EIA.)*

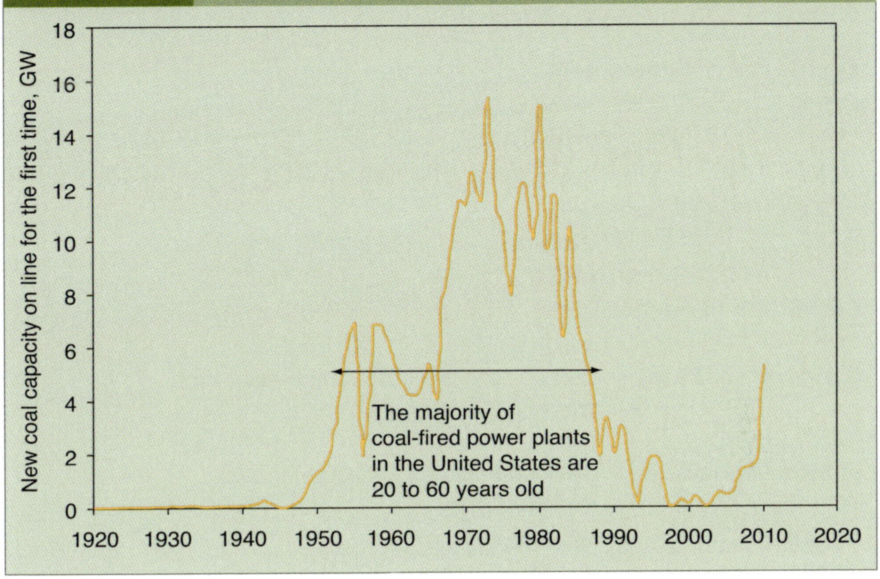

because, after injection ceases, the site will need to be monitored to ensure that the injected CO_2 is behaving as expected.

The sections that follow describe capture and transport technology and the life cycle of a geologic sequestration project in more detail, focusing particularly on issues that should be considered in the context of developing a regulatory framework for CO_2 capture, transport, and geologic sequestration.

2.1 OVERVIEW OF CO_2 CAPTURE TECHNOLOGY AND ITS APPLICATIONS

Carbon dioxide can be captured from a large number of industrial processes, including electricity-generating plants, natural gas processing, chemicals production, and cement manufacturing. Systems are also being developed to capture CO_2 directly from the air.[3] Regardless of the process, the economics of CO_2 capture and transport exhibit returns to scale; that is, the unit cost of capturing CO_2 decreases with increasing capture volume. Thus, large point sources of CO_2 emissions will be the most attractive sources for CO_2 capture. Typical features of various large point sources of CO_2 are shown in Table 2.1.

TABLE 2.1 *Typical direct CO_2 emissions rates from large point sources of CO_2.*[a]

Process	CO_2 Source	Size	CO_2 Emissions (Mt/y)
Electric Power Generation[b]			
NGCC	Combustion—natural gas	500 MWe	1.2
SCPC	Combustion—coal	500 MWe	2.7
IGCC	Gasification—coal	500 MWe	2.7
Chemical & Petrochemical			
CTL–Synthetic Fuels	Gasification & Fischer–Tropch synthesis	50,000 bbl/day	11
Ethylene Production	Thermal cracking of light hydrocarbons	700,000 tonnes/y	1.7
Ammonia Production	Haber–Bosch process	500,000 tonnes/y	0.6
Hydrogen Production	Steam-methane reforming	70,000 tonnes/y	0.3
Other Industrial			
Integrated Steel Production	Blast furnace emission	3,000,000 tonnes/y	3.8
Cement Production	Clinker production	700,000 tonnes/y	0.6

Notes: a. Actual emissions rates will vary from these numbers depending on the specific feedstocks and processes employed and actual greenhouse gas emissions to the atmosphere will be higher on a life cycle basis. Estimates for electric power generation come from the Integrated Environmental Control Model (IECM),[1] for coal-to-liquids (CTL) synthetic fuels from the National Academies Report,[2] and those for chemical, petrochemical, and other industrial processes from the IEA Greenhouse Gas Programme.[3]

b. A plant capacity factor of 75% is used for each of the power plants; NGCC, SCPC, and IGCC refer to Natural Gas Combined Cycle, Supercritical Pulverized Coal, and Integrated Gasification Combined Cycle plants, respectively.

Sources: 1. Rubin, E. S., C. Chen, et al. 2007. Cost and performance of fossil fuel power plants with CO_2 capture and storage. *Energy Policy* 35(9): 4444–4454.

2. National Academy of Sciences, National Academy of Engineering, et al. 2009. *Liquid Transportation Fuels from Coal and Biomass: Technological Status, Costs, and Environmental Impacts*. Washington, DC: National Academy of Sciences.

3. Hendriks, C., A. S. van der Waart, et al. 2002. *Building the Cost Curves for CO_2 Storage, Part 1: Sources of CO_2*. Stoke Orchard, UK: IEA Greenhouse Gas R&D Programme.

2.2 CAPTURE OF CO_2 FROM ELECTRIC POWER GENERATION

Carbon dioxide can be captured from electric power generation using three different methods: 1) pre-combustion CO_2 capture, in which carbon is removed from the fuel prior to oxidization; 2) post-combustion CO_2 capture, in which fuel is combusted normally in a boiler or turbine and CO_2 is then removed from the exiting flue gas stream; and 3) oxyfuel combustion, in which the fuel is combusted with only the amount of oxygen necessary for complete combustion in an atmosphere of CO_2. Figure 2.2 summarizes the major flows and processes in the three categories of capture processes.

Pre-combustion Capture

Pre-combustion CO_2 capture is characterized by removal of the carbon from fuel prior to combustion in a gas turbine. If the starting fuel is a solid, such as coal, petroleum coke,[4] or biomass (e.g., wood chips, switchgrass, etc.), the fuel is converted to a gas through partial oxidization of carbon in the fuel in the presence of oxygen. The resulting gas, referred to as "syngas," consists primarily of hydrogen

FIGURE 2.2 *CO_2 capture processes in electric power generation summarized by major process steps, where the thick lines show the path of carbon through the processes. (Figure adapted from IPCC Special Report on Carbon Dioxide Capture and Storage.)*

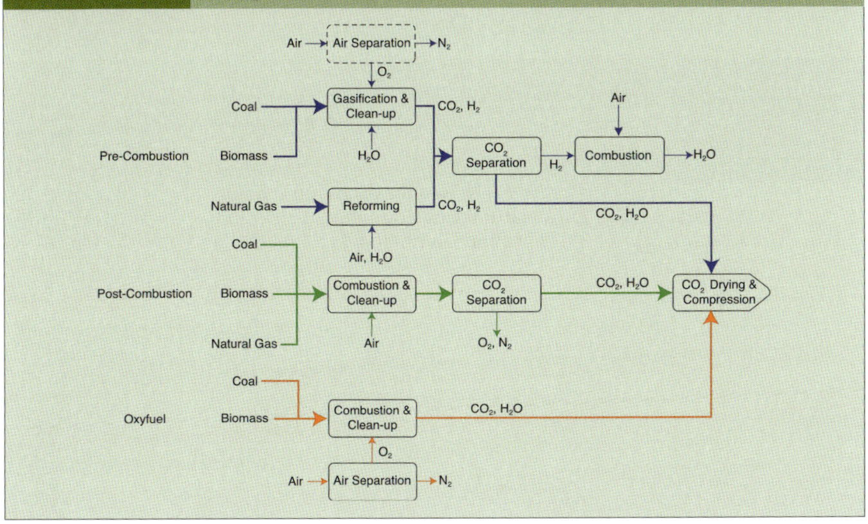

(H_2), carbon monoxide (CO), and light hydrocarbons—typically methane (CH_4), although the predominant light hydrocarbon depends on the gasifier design.[5] At this point, the gas can be treated to remove hydrogen sulfide (H_2S) and fine particulates, and then burned in a combined cycle gas turbine to generate electricity.[6] This process is referred to as the Integrated Gasification Combined Cycle (IGCC) process. There are a number of IGCC power plants operating, located in the US, Europe, and Japan (see Box 2.1). In addition, there are hundreds of such coal gasifiers operating around the world providing feedstock to chemical plants.

In an IGCC plant with CO_2 capture, the hydrogen content of the synthesis gas is increased by reacting CO in the synthesis gas with water through the water gas shift reaction, as shown in Figure 2.3. In this reaction, CO is converted to CO_2, which is then removed from the fuel using a separation process. Prior to CO_2 separation, H_2S produced during gasification will typically be removed and converted to elemental sulfur, although if regulations were to allow, it could be co-captured and sequestered along with CO_2.

Processes to separate CO_2 from the gas stream include:

- physical absorption of the CO_2 into a solvent which is then regenerated through pressure and temperature changes;
- adsorption of the CO_2 onto a solid that is periodically regenerated, releasing the CO_2 (pressure or temperature swing adsorption); and
- transfer of the CO_2 across a thin membrane that is permeable only to CO_2.

Currently, for IGCC systems, the preferred separation method is physical absorption using a solvent because this technology is commercially available and can be scaled to handle large volumes of CO_2 most easily.[7] For production of pure H_2, solid sorbents are preferred, because they yield a higher purity product. Membrane systems are largely under development at the laboratory or bench scale.[8]

FIGURE 2.3 *An expanded view of the pre-combustion CO_2 capture processes for electricity generation, where the thick lines show the flow of carbon through the process.*

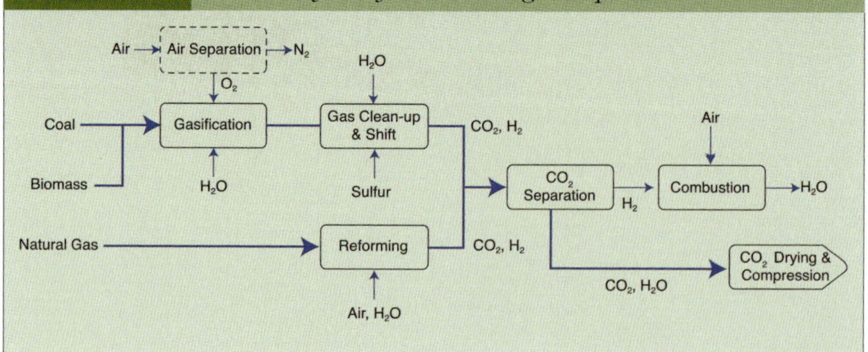

Box 2.1 *Demonstrations of CO_2 Capture from Power Generation*

At the time this book was written, there are approximately a dozen operating pilot projects capturing CO_2 from power plants. These power plants range in size from less than a megawatt to several tens of megawatts.[9] In addition, there are several other plants in commercial operation capturing CO_2 from a small fraction of the flue gas from power plants, the CO_2 from which is purified and sold for use in food processing (e.g., beverage carbonation).

The three largest of the operating pilot plants capture approximately 0.1 million tonnes (Mt) of CO_2 per year, or the emissions equivalent to those from a 10 MW capacity power plant. Two of these projects—the Schwarze Pumpe Pilot Plant near Spremberg, Germany, and the Lacq Pilot Project in Lacq, France, respectively—capture CO_2 using the oxyfuel combustion process.

Schwarze Pumpe 30 MW thermal CCS oxyfuel pilot plant in Spremberg, Germany. (Photo courtesy of Vattenfall.)

Lacq 30 MW thermal CCS oxyfuel pilot plant in Lacq, France. (Photo courtesy of Total.)

The third project, the Mountaineer CCS Project near New Haven, West Virginia, captures CO_2 using a chilled ammonia post-combustion capture process. After successfully operating, this pilot plant was shut down in the summer of 2011.

Chilled ammonia CCS demonstration plant at the Mountaineer Power Plant outside of New Haven, West Virginia. (Photo courtesy of American Electric Power.)

There are nearly twenty operating IGCC plants worldwide.[10] Two of these in the US that are now generating electricity are located in Polk County, Florida, and near West Terre Haute, Indiana.

The Polk IGCC demonstration plant in Polk County, Florida. (Photo courtesy of USDOE.)

The Wabash River IGCC plant near West Terre Haute, Indiana. (Photo courtesy of USDOE.)

There are plans to build a number of IGCC facilities in the US that would capture CO_2 for sequestration on a large scale. Most of these US projects hope to sell the captured CO_2 for use in Enhanced Oil Recovery (EOR).[11] At the time of writing, a megawatt scale pilot plant is in operation at an IGCC plant near Puertollano, Spain.

If natural gas is used as the fuel for a pre-combustion capture process, the gas (which is primarily CH_4) is converted to H_2 and CO_2 through steam-methane reforming, partial oxidization, or some combination of the two processes (known as autothermal reforming).[12] In this process, CH_4 and steam are converted to H_2 and CO which are shifted to H_2 and CO_2 using the water–gas shift reaction. The subsequent CO_2 separation and combustion processes are the same as in the IGCC case, save for H_2S removal, since the sulfur content of sweetened natural gas is typically very low.

Post-combustion Capture

CO_2 capture using these methods involves combustion of fossil fuels in the traditional manner (i.e., using a boiler or a combined cycle gas turbine) and subsequent separation of CO_2 from the flue gas. While several processes are capable of separating CO_2 from other flue gas constituents, the current preferred option is absorption using a chemical sorbent such as amines or ammonia (Figure 2.4). Because of the relative simplicity of post-combustion capture processes, they can be applied to the flue gas from virtually any combustible fuel, such as natural gas, coal, or biomass. However, where solid fuels, such as coal or biomass, are used, concentrations of acid gases like SO_x or NO_x in the flue gas must be substantially reduced to prevent degradation of the solvent.[13]

FIGURE 2.4 *A flow diagram showing the flow of carbon in post-combustion CO_2 capture. Flue gas clean-up is only required for solid fuels, as natural gas contains low concentrations of sulfur and combustion can be controlled to minimize the concentration of NO_x.*

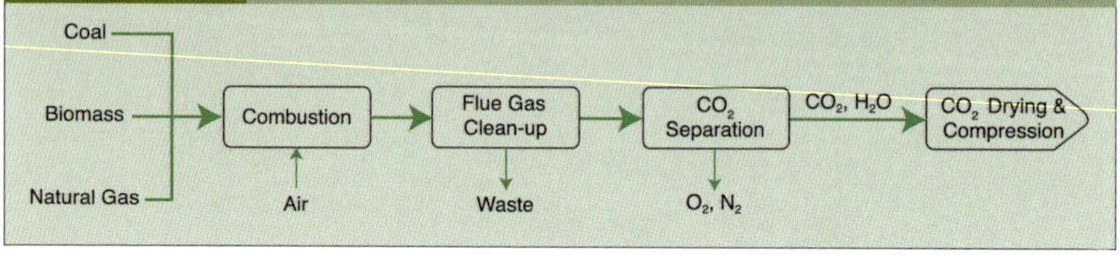

Oxyfuel Capture

The third option for CO_2 capture is the combustion of biomass or coal in an atmosphere of only oxygen and CO_2, rather than in air (Figure 2.5).[14] In principle, the only products of combustion are water and CO_2, thus no CO_2 separation is required. The advantages of oxyfuel combustion relative to post-combustion capture result from the absence of CO_2 separation, and reduced volumes of flue gas that must be handled—this latter factor resulting in smaller equipment sizes.[15] However, an air separation unit is needed to produce oxygen, and care must be taken to avoid system leakage points that can allow nitrogen to enter the process.

To control the temperature of combustion in the boiler, CO_2-rich flue gas produced from combustion is recycled back into the boiler. In practice, there are also small amounts of conventional pollutants, such as SO_x, NO_x, and fly ash, in the flue gas.[16] Prior to recycling part of the flue gas stream, SO_x and fly ash are typically removed to make handling the flue gas easier and control the concentration of these pollutants in the CO_2 that is compressed for sequestration.[17]

Compression and liquefaction of the captured CO_2 is the final step in all of the capture processes. During this step, the concentration of water in the CO_2 must be reduced to avoid formation of CO_2 hydrates and to avoid corrosion problems in

FIGURE 2.5 *A flow diagram showing the flow of carbon in oxyfuel combustion.*

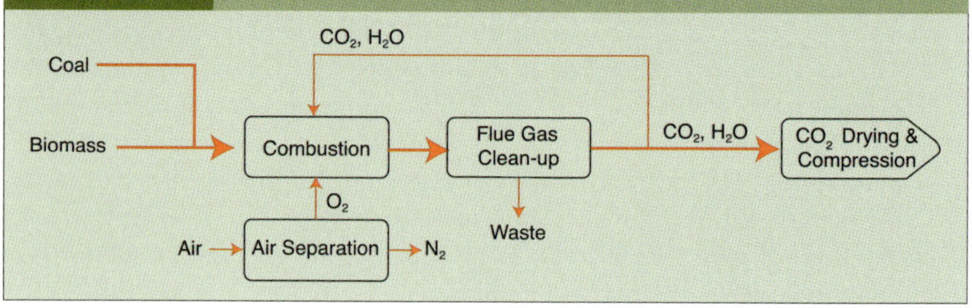

subsequent transport. In addition, the concentration of gaseous components such as N_2, O_2, and Ar may have to be reduced, particularly if the CO_2 is used for EOR applications.[18]

Energy Penalty Impacts and the Cost of Capture

All of these types of capture systems require large amounts of energy for their operation, resulting in decreased plant efficiencies and reduced net power outputs when compared to the same plants without capture systems. Because of the energy consumption for CO_2 separation and compression, the emissions of CO_2 avoided by capturing CO_2 are always less than the amount of CO_2 captured. This concept is illustrated in Figure 2.6.

In addition, reduced efficiency of electric power generation means that power plants with capture systems will consume more fuel and water while producing more waste products (e.g., ash, slag, and sulfur) per unit of electricity generated than a plant without capture.[19] This increase in resource consumption can be quantified by the energy penalty, which is defined as the increase in plant energy input per unit of product or output.[20] Table 2.2 shows that the energy penalty ranges from roughly 17% to 33% depending on the plant type. The results reported in this table agree broadly with those from other recent studies.

It is important to note that the amount of CO_2 avoided based on the results in Table 2.2 accounts only for the avoided direct emissions from power generation. If the boundary of the capture system is increased to include second order emissions from the coal supply chain, sorbent production chain, and the production chain of reagents for other environmental control systems, the emissions reduction is smaller than the emissions avoided by capture. For example, Koornneef et al. have

FIGURE 2.6 *Illustration showing the relative difference between CO_2 captured and emissions avoided. (Figure courtesy of Herzog and Smekens, IPCC Special Report on Carbon Dioxide Capture and Storage.)*

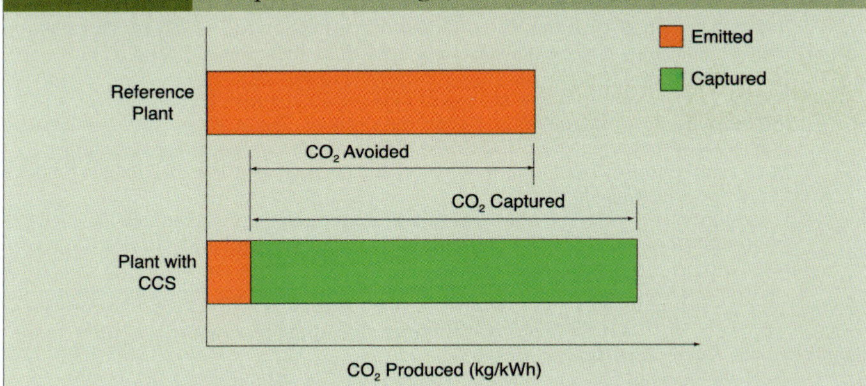

Table 2.2 *Key assumptions and baseline results from Finkenrath.*[a]

Parameter	PC[b]		IGCC[c]		NGCC[d]	
	Reference	Capture	Reference	Capture	Reference	Capture
Net power output (MW)	582	545	663	546	528	461
Net Efficiency, LHV (%)	41.4	30.9	41.4	33.1	56.6	48.4
CO_2 Emission Rate (kg/MWh)	820	111	793	115	370	55
Decrease in Net Efficiency (%)		25		20		15
Overnight Cost ($/kW)[e]	2162	3308	2586	3714	960	1715
LCOE ($/MWh)	66	107	75	104	77	102
Cost of CO_2 avoided ($/tonne CO_2)[f]		58		43		80

Notes: a. Costs are for capture (including compression) and LCOE is calculated with a 10% real discount rate. All costs in constant 2010 US $ and *do not include* transport and storage.

b. Results are averages from a re-analysis of thirteen previous studies conducted by the IEA. All of the original studies were of post-combustion capture systems with amine-based solvents. The average capture efficiency and capacity factor across all studies was 87% and 83%, respectively.

c. Results are averages from a re-analysis of eleven previous studies conducted by the IEA. The original studies use a variety of gasification technologies. The average capture efficiency and capacity factor across all studies was 88% and 81%, respectively.

d. Results are averages from a re-analysis of nine previous studies conducted by the IEA. All of the original studies use F-Class gas turbines. The average capture efficiency and capacity factor across all studies was 87% and 88%, respectively. Natural gas price assumptions depend on the region of the underlying study: $9.76/GJ in OECD Europe and $7.40/GJ in the US.

e. Overnight costs include owner's, EPC, and contingency costs, but not IDC. A 15% contingency based on EPC cost is added for CCS cases and a 5% to non-CCS cases.

f. All avoided cost values are relative to the reference plant for the same system.

Source: Finkenrath, M. 2011. Cost and performance of carbon dioxide capture from power generation. *IEA Energy Papers*. No. 2011/05.

estimated that, including second order emissions, the emissions reduction for a pulverized coal-fired power plant with post-combustion capture will be 71%, as opposed to the 87% for just the plant itself.[21] Of course, the second order emissions could be reduced by increasing the efficiency of the capture process, which would result in fewer impacts from coal mining and transport, and reducing emissions from coal mining and transport. Discussion on accounting for emissions reductions from CO_2 capture can be found in Chapter 8.

The decrease in energy output coupled with the increased operating cost and increased capital cost (resulting from the capture system) results in a higher levelized cost of electricity (LCOE) from plants capturing CO_2 than for those without CO_2 capture systems. Table 2.2 shows that, with current designs, the cost of electricity from a power plant capturing CO_2 can be expected to increase between 30% and 70% relative to the reference plant, but as noted below, this can be expected to decline as experience is gained with more plants.

The technological maturity of CO_2 capture systems varies from technology to technology. The 2005 report from the Intergovernmental Panel on Climate Change (IPCC) concluded that the components of many capture systems are well understood from a technological standpoint, but there is a lack of experience in building and operating capture systems at scale in the electric utility industry.[22] More recent work on learning curves associated with technologies similar to CCS by Rubin and his coworkers (see Chapter 9) suggests that costs may increase for the first several plants as design and operating strategies are refined.[23] However, once this period is past, the cost of energy from power plants with CO_2 capture can be expected to decrease by between 3% and 5% for each doubling of worldwide installed capacity.[24] Worldwide capacity is used as the metric for learning, implying that the barriers for technology and knowledge transfer are relatively low. However, if a nation were to develop CCS technology domestically, domestic contractors and operators could gain a considerable competitive advantage in being able to build and operate these plants. For example, CCS technology appears particularly important to China if the country is to reduce emissions,[25] and some have speculated that rapid development of indigenous capture technology in China is being driven in part by the anticipation of export opportunities.[26]

2.3 CAPTURING CO_2 FROM INDUSTRIAL PROCESSES

The ease with which CO_2 can be captured from industrial processes is highly dependent on the nature of the process. Capture from some processes, such as gas cracking to produce ethylene, gas processing, and coal-to-liquids (CTL) is relatively simple and low cost because CO_2 separation is already a necessary part of the process. For others, such as cement production, capture will be more complex and costly.

Processes can be grouped into those that produce a stream of relatively pure CO_2 and currently vent it to the atmosphere, those that do not, and those that produce CO_2 from both combustion and chemical reactions. Table 2.3 describes the CO_2 source and capture processes for these industrial processes.

Table 2.3 *Summary of CO_2 capture options from industrial processes.*

Process	CO_2 Source	Technology	Relative Capture Cost[a]
Inherent CO_2 separation			
Gas Processing[1]	Natural gas commonly contains CO_2 and other components that are removed so that the gas meets pipeline specifications (i.e., approximately 2% CO_2)	■ Physical and chemical solvents (e.g., Selexol, MDEA) ■ Membranes ■ Phase change (e.g., ExxonMobil CFZ)	Low—Clean-up and compression only.
CTL-Synthetic Fuels[1]	Syngas, produced from gasification, is used to produce liquid fuels. CO_2 is separated from the syngas both prior and after conversion to liquid fuels.	■ Physical and chemical solvents (e.g., Selexol, MDEA) ■ Membranes	Low—Clean-up and compression only.
Ammonia Production[1]	In the US, ammonia is typically produced from steam reforming of natural gas with a subsequent shift to produce H_2 and CO_2. The CO_2 is normally separated from the hydrogen, which is then combined with N_2 to produce ammonia. CO_2 from ammonia synthesis is sold for Enhanced Oil Recovery in some cases.	■ Physical and chemical solvents (e.g., Selexol, MDEA) ■ Membranes	Low—Clean-up and compression only.
CO_2 from multiple sources			
Hydrogen Production[1]	Natural gas can be converted to hydrogen through steam-methane reforming (SMR), partial oxidization (POX), and autothermal reforming (ATR). In all of these processes, CO_2 is separated from H_2 in normal operation; however, for SMR and POX current processes will require additional separation.	■ Physical solvents (e.g., Selexol, MDEA) ■ Membranes ■ Pressure-swing adsorption	Comparable—May require addition of flue-gas separation.

Ethylene Production	Ethylene is typically produced from the thermal cracking and reforming of some combination of ethane, propane, butane, and naptha. CO_2 is separated from the produced stream of ethylene to meet quality requirements; however, CO_2 is not normally captured from fired heaters.	■ Physical and chemical solvents (e.g., Selexol, MDEA) ■ Membranes	Comparable—May require addition of flue-gas separation.
Cement Production[2]	CO_2 is generated directly through calcination as well as from combustion of fossil fuels to produce heat for the calcining reaction.	■ Physical and chemical solvents (e.g., Selexol, MDEA) ■ Oxyfuel combustion	High—Similar processes are not in use at cement plants today.
Integrated Iron and Steel[3]	In traditional processes, CO_2 is released from coke production, and combustion in a blast furnace. Some more recent processes forgo the use of coke and use coal directly and are more amenable to CO_2 capture.	■ Oxyfuel combustion for blast furnaces ■ Direct reduction of iron ore by hydrogen derived from gasification	Unknown—Would require significant addition of equipment and changes in operation.

Note: a. Relative to the cost of CO_2 capture of $36.5/tonne for the PC plant with capture described in Section 2.3 not including transport and storage.

Sources: 1. Thambimuthu, K., M. Soltanieh, et al. 2005. Capture of CO_2. In *IPCC Special Report on Carbon Dioxide Capture and Storage*. B. Metz, O. Davidson, H. de Coninck, M. Loos, and L. Meyer, eds. Cambridge: Cambridge University Press.

2. Hassan, S. M. N., P. L. Douglas, et al. 2007. Techno-economic study of CO_2 capture from an existing cement plant using MEA scrubbing. *International Journal of Green Energy* 4(2): 197–220.

3. Gielen, D. 2003. CO_2 removal in the iron and steel industry. *Energy Conversion and Management* 44(7): 1027–1037.

It is important to note that of the four integrated, commercial-scale CCS projects operating today, three capture CO_2 from natural gas processing and one from synthetic natural gas production. There are also several facilities in the US, primarily ammonia production and gas processing, that capture CO_2 for use in enhanced oil recovery.

2.4 CAPTURING CO_2 DIRECTLY FROM THE AIR

At present, there are two technically feasible approaches by which CO_2 can be captured from the air: combustion of biomass with capture of CO_2 to generate electricity,[27] and direct removal of CO_2 from the atmosphere.[28]

Because biomass removes CO_2 from the atmosphere during growth, converting it into various biological molecules (e.g., carbohydrates, lipids, and proteins) and oxygen through photosynthesis, biomass can be used as a fuel source in industrial facilities that employ CCS. Biomass combustion to generate electricity is an attractive option because not only does it result in negative emissions, but the electricity generated could be used to displace that generated from fossil fuels, resulting in large avoided emissions.[29] However, use of biomass for energy is constrained by supply considerations. For example, widespread use of food crops (e.g., corn, soybeans, rapeseed, etc.) for energy has already resulted in direct competition between energy and food supplies and, even when non-food crops are used (e.g., corn stover, switchgrass, etc.), indirect impacts on food prices and emissions could occur due to changes in land use patterns.[30]

There is interest in direct capture of CO_2 from the air because the cost of this technology presents an upper bound (or "backstop") on the cost of removing carbon from all sectors of the economy.[31] With a suitable supply of low carbon energy, air capture could potentially remove more CO_2 from the atmosphere than is emitted from human activities. As a result, air capture may afford society the option of directly reducing atmospheric concentrations of CO_2.[32]Viable air capture technology could be particularly attractive as a strategy for offsetting emissions from small and mobile sources such as automobiles and aircraft.[33]

Direct air capture can be performed using a number of different chemical processes and even systems based on the same chemistry may have subtle differences.[34] One process that has been analyzed from both an engineering and an economic standpoint is shown in Figure 2.7. Figure 2.8 is an artist's rendering of what such a capture system might look like.

In this process, CO_2-rich ambient air is contacted with sodium hydroxide, NaOH, in a spray tower[35] or packed column,[36] producing sodium carbonate, Na_2CO_3. Sodium carbonate is then converted to calcium carbonate, $CaCO_3$, by reaction with calcium hydroxide, $Ca(OH)_2$. Calcium carbonate is decomposed in a kiln at approximately 900°C, producing lime (CaO) and gaseous CO_2. Flue gas from the kiln can then be treated by a post-combustion capture process using amines or another physical solvent.[37] Alternatively, the kiln could be operated in

FIGURE 2.7 *The layout of the direct capture system where thick arrows represent flows of carbon in the process.*

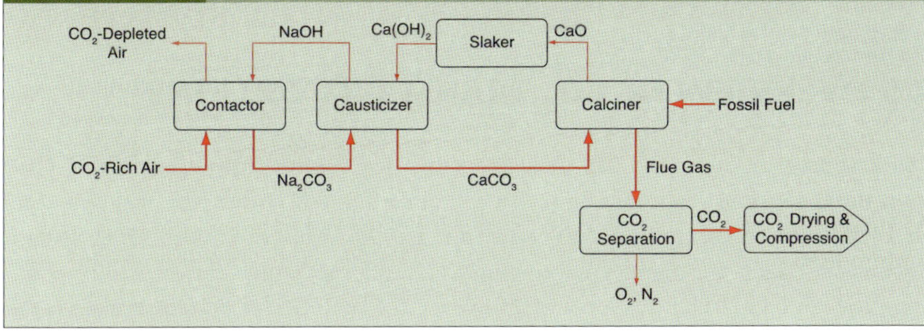

FIGURE 2.8 *Artist's rendering of an air-capture unit, designed by Carbon Engineering Ltd., to capture about 100 k-tonnes of CO_2 per year. A battery of such air-capture units would work in unison with a chemical recovery plant to produce high purity CO_2 that has been captured directly from air. (Reproduced with permission from Carbon Engineering.)*

an oxyfuel mode with the addition of air separation to produce oxygen and the removal of CO_2 separation.[38]

Depending on the design of the contactor and downstream process, total energy requirements for this process (including compression) appear to be between 13 and 18 GJ per tonne of CO_2 captured.[39] For comparison, coal and gas contain approximately 10 and 20 GJ of energy per tonne of CO_2 produced. Thus, with current technology, emissions reductions could be achieved only if low-carbon energy sources (i.e., with a carbon intensity equal to or less than that of natural gas without CCS) are available. Alternative process designs being developed today could reduce the energy requirements of regenerating the sodium hydroxide (NaOH) solution,[40] and other processes could reduce energy through use of novel solid sorbets,[41] or the direct use of solar energy.[42]

2.5 OVERVIEW OF CO_2 TRANSPORT OPTIONS

Pipeline, trucks, ships, and rail are all options for transporting compressed CO_2 from the location of capture to the sequestration site. The most practical and cost-effective method of transport depends on the locations of capture and storage, distance between the facility capturing CO_2 and the sequestration site, and the quantities of CO_2 to be transported. However, the quantity to be transported is the dominant factor—as shown in Table 2.1, on the order of 2 to 3 Mt per year of CO_2 would be produced from a single 500 MW coal-fired power plant capturing CO_2. As a result, pipeline is the only viable option for overland transport from power plants,[43] and is the only method of transport that is considered in this work.

In the US, there is considerable industrial experience in the transport of CO_2 by pipeline. In 2009, nearly 60 Mt of CO_2 was transported over approximately 5,800 km of pipelines primarily for use in CO_2-flood enhanced oil recovery (EOR) operations. Even without CCS, both the mass transported and the scale of the network is expected to grow as a result of growing demand for CO_2 for use in EOR.[44] Over 80% of this CO_2 is produced from natural accumulations, such as Jackson Dome (Mississippi), McElmo Dome (Colorado), and Bravo Dome (New Mexico).[45] For comparison, 60 Mt of CO_2 would be the amount of CO_2 produced by nearly twenty 500 MW coal-fired power plants annually.

From an operational perspective, pipeline diameters are sized according to operating parameters so that CO_2 remains supercritical fluid, as shown in Figure 2.9, throughout transport.[46] CO_2 pipeline diameters vary, but generally larger diameters of pipe result in lower transportation costs.

The cost of CO_2 transport is much lower than the cost of CO_2 capture from electric power generation (Table 2.2) or industrial sources (Table 2.3). The cost of CO_2 transport by pipeline decreases exponentially with the mass of CO_2 transported, increases approximately linearly with distance, and is sensitive to the terrain through which the pipeline passes. Terrain is important not only because it affects design and construction of the pipeline, but also because it affects the costs of acquiring the right-of-way for the pipeline. The IPCC estimates the average cost of CO_2 transport to be on the order of dollars per tonne of CO_2 for pipelines transporting millions of tonnes of CO_2 per year over hundreds of kilometers,[47] which has been supported by more recent work.[48] The initial capital cost of such a pipeline system would typically be around \$1 million per kilometer.

It also appears that CCS will not drastically alter current siting calculations for electricity generation units, as the "cost of piping CO_2 is not negligible, but much less than [electric] transmission cost."[49]

FIGURE 2.9 *Pressure-temperature phase diagram for carbon dioxide (CO_2). At pressures greater than 7.38 Mpa and temperatures greater than 31°C (known as the critical pressure and temperature, respectively), CO_2 is a supercritical fluid.*

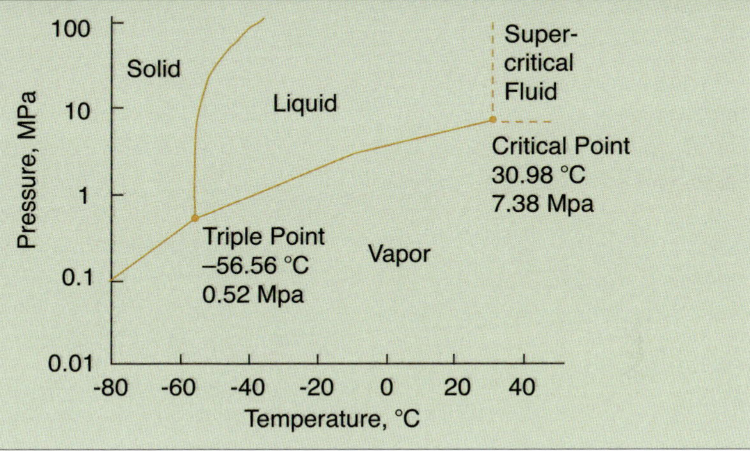

Many studies have made estimates of the future amounts of CO_2 that could be captured through CCS technology both in the US and worldwide.[50] However, few of these studies explicitly address the scale of pipeline infrastructure required and, among those studies that do, the differences are large. For the US, estimates range from less than 18,000 km of new pipelines built over a forty-year period,[51] to more than 58,000 km over a similar period.[52] Estimates of the future scale of a CO_2 pipeline network are fraught with uncertainty not only due to questions about the future stringency of emissions constraints and the role that CCS will play in meeting these constraints, but also due to uncertainty over the relative locations of facilities that may capture CO_2 and suitable sequestration sites, economics of geologic sequestration, and the role that enhanced oil recovery will play in development of such a system.[53] However, it is clear that the economies of scale dictate that a networked infrastructure will result in lower transport costs than point-to-point pipelines.[54]

These uncertainties notwithstanding, the development of a pipeline network that transports hundreds of millions to billions of tonnes of CO_2 over tens of thousands of kilometers of pipelines in the US is technically and economically feasible. As a point of comparison, US consumers used approximately 500 Mt of natural gas in 2010, the vast majority of which was delivered by pipeline. The current US gas transmission system includes approximately 480,000 km of pipelines (Figure 2.10). As shown in Figure 2.11, approximately 40% of the mileage—190,000 km—of the gas transmission system was built over the past forty years.[55]

FIGURE 2.10 *US gas pipeline system as of 2009. Blue indicates interstate pipelines; red indicates intrastate pipelines. (Figure from US EIA.)*

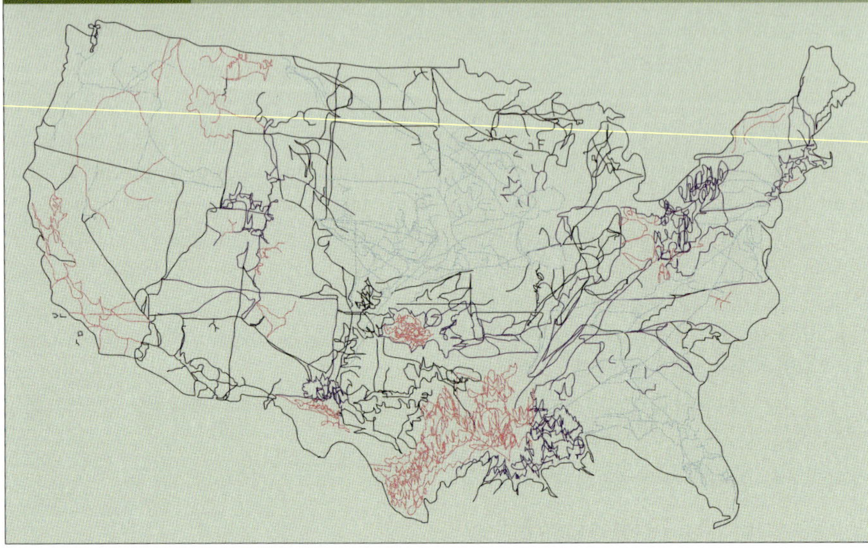

FIGURE 2.11 *Total mileage of the US natural gas transmission system by decade since 1940 (solid line), as well as incremental additions to the network each decade since 1940 (bars).*

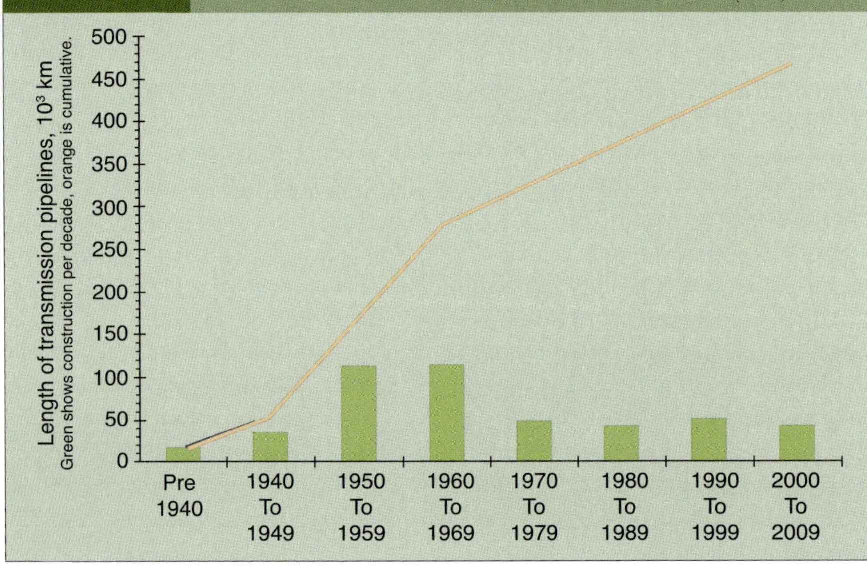

2.6 OVERVIEW OF GEOLOGIC SEQUESTRATION AND THE SEQUESTRATION PROJECT LIFE CYCLE

The life cycle of an onshore geologic sequestration project is illustrated in Figure 1.8 in the previous chapter. We will refer to this diagram from time to time in later chapters as we address issues that arise in its various stages.

While the focus of this book is legal and regulatory barriers to sequestration in geologic formations located onshore in the US, offshore sequestration represents the vast majority of the sequestration resource for some nations (e.g., the United Kingdom, Norway), and two of the first large-scale sequestration projects inject CO_2 into formations offshore. Sequestration under the deep seabed (i.e., approximately 2 km of water) may also be attractive because the density of CO_2 approaches the density of sea water at such depths, reducing or eliminating risks of leakage; approaches to pressure management may be simpler and less expensive; and some mineralization reactions may be accelerated.[56] The life cycle for an offshore project and legal and regulatory barriers faced by such a project may be somewhat different, and are not explicitly addressed in this book.

Exploration, Screening, and Characterization

Through the Regional Carbon Sequestration Partnerships, the US Department of Energy (DOE) compiles high-level estimates of the sequestration resource for the United States. Most recently, the DOE estimates the volume of accessible pore space as sufficient to sequester between 1,100 and 13,500 Gt of CO_2 onshore in the continental US.[57] Substantial additional resources—on the order of 10^3 Gt—may be available under the sea floor in state and federal waters.[58] Assessments at larger scales and higher resolutions (e.g., basin- and region-scale assessments) are a necessary next step to identify possible storage formations and the portion of the resource that is economic storage capacity.[59]

During the site screening phase, locations of CO_2 sources, information on regional geology (e.g., regional cross-sections, outcrop analysis), and publicly available site-specific data from oil and gas exploration (e.g., seismic data, well logs, core analysis) must be analyzed to identify candidate sites. This analysis should focus on the suitability of the regional geology for sequestration (e.g., the presence of a regional sealing formation, absence of extensive faulting and fracturing, etc.) and a screening level economic analysis (e.g., existence of sufficient capacity and adequate injectivity, permission to access pore space, etc.).[60] The screening assessment may also need to consider the potential for interactions with other sequestration projects in the region, as regional pressure changes may impact the effective capacity of the site.[61]

The next step after a screening assessment is the more detailed characterization of site-specific geology in the most promising sites. Site characterization will include geophysical methods such as seismic and wireline logging, as well as

Box 2.2 *Site Characterization Methods*

A variety of tools have been developed for oil and gas exploration and production and groundwater remediation that are applicable to exploration for CO_2 storage sites. These tools include:[62]

Seismic Surveying: Seismic waves are generated from an energy source at the surface (such as small explosive charges and vibroseis trucks on land, or air guns in water) and an array of detectors that record energy that returns to the surface due to reflection (or refraction) with subsurface geological features. The travel time of the seismic waves that arrive at the detectors can then be used to estimate the depth to geologic features of interest. In two-dimensional (2-D) seismic, data are gathered sequentially along linear seismic lines, while in three-dimensional (3-D) seismic, data are typically gathered by placing detectors along "in-lines" orthogonal to a line of "shot-points," also referred to as "cross-lines." Although 3-D seismic is considerably more expensive than 2-D seismic, it has the benefit of generating higher resolution data that can be used to image a volume of the subsurface rather than only a vertical plane, as in 2-D seismic. Seismic receivers or energy sources can also be placed in a borehole in Vertical Seismic Profiling (VSP), which, when coupled with traditional surface seismic, may allow ambiguities in the data to be resolved.

Geophysical Well Logging: A tool or "sonde" is lowered to the bottom of a borehole and records information about the surrounding rock and fluids as it is drawn up the well. A large number of geological properties can be recorded using logging tools, including formation thickness and lithology, porosity, permeability, fluid saturation, and formation dip. Typically, a number of tools are combined to allow a number of properties to be estimated on one run. In addition, some well logging tools can be incorporated into the drill string to log-while-drilling (LWD).

Core Sample Analysis: During drilling, a specialized bit can be used to retrieve a largely intact core of rock from the borehole. A large number of properties can then be measured in the lab, including relative permeability curves, seal capillary entry pressure, and rock strength.

Pressure Testing: The pressure at the bottom of the well is decreased by producing fluids from the well. Once steady state is reached, the production is stopped and the well is shut in. Measurement of the gradual return of pressure to equilibrium allows the hydraulic properties of the reservoir to be estimated.

geologic interpretation and evaluation supplemented by analysis of core and fluid samples from exploration wells (see Box 2.2). Potential sequestration sites must be shown to have sufficient capacity and injectivity to accept the desired volumes of CO_2 injected at desired rates, and geology that will effectively contain CO_2 in the long term. The information needed to make this case will likely include:

- structure contour maps of the storage unit, sealing units (primary and secondary), and overlying aquifers;

- geologic maps showing any lateral changes in the storage or sealing unit lithology;
- storage and primary sealing unit petrophysical (e.g., porosity, permeability) and geochemical (i.e., mineralogy) properties;
- reservoir pressure and temperature conditions;
- fluid properties of both the injected CO_2 and the pre-existing reservoir fluids in the storage unit and overlying aquifers;
- magnitude and direction of water flow, hydraulic connectivity of formations, and pressure changes associated with hydrocarbon production—the latter if applicable;
- evaluation of natural seismicity and tectonic activity of the region and characterization of the in-situ stress regime in the storage unit and primary seal;
- determination of rock strength properties, deformation properties, thermal properties and attributes of weak planes and discontinuities of the storage and primary sealing units;
- documentation and maps of faults and fractures in both the storage and the sealing units, including their stability under injection pressures;
- oil and gas production data (if the storage unit is an oil or gas reservoir);
- locations of nearby wells and assessment of their integrity;
- maps of the predicted pathway along which the CO_2 will migrate from the point of injection;
- detailed maps of the structural boundaries of traps where the CO_2 is expected to accumulate, especially highlighting potential spill points;
- assessment of the pressure build-up during injection; and
- assessment of the geochemical stability of both the primary seal and the storage unit.

Data generated from the application of geological, geophysical, and reservoir engineering tools, such as those discussed in Box 2.2, allow a detailed model of the target reservoir to be developed. The results from this modeling can be used to project the fate of the injected CO_2 over the long term, to design the placement and configuration of injection wells,[63] and to devise monitoring plans. Pilot injection of CO_2 (or other suitable fluids) will greatly further understanding of the reservoir properties and the site-specific behavior of the injected CO_2, allowing the geological model to be refined.[64] Screening analysis, site characterization and evaluation, and pilot injection and monitoring are collectively referred to as "site screening and characterization" in this book.

The site screening and characterization process is perhaps the most important step in the development of a sequestration project. The oft-cited IPCC conclusions—that the fraction of injected CO_2 retained by sequestration projects "is very likely" to be "more than 99% over the first 1000 years" and that risks to human health and the local environment are no greater than those associated with similar oil and gas industry operations—are contingent on sites being both "well-characterized" and "well selected, designed, operated and appropriately monitored."[65] The corollary to this is that no measure of good design, operation, and monitoring will prevent unintended migration and, possibly, leakage of injected CO_2—with the attendant risks to human health and the environment—from a site that is poorly suited for CO_2 sequestration. Best practices guidelines for site screening and characterization have been or are being developed by several different organizations, including the World Resources Institute (WRI),[66] the US Department of Energy National Energy Technology Laboratory,[67] Det Norske Veritas,[68] and the Canadian Standards Association.

To date, these best practice guidelines have not incorporated the two-way dialog that will be necessary to gain public acceptance of suitable sites.[69] Once screening and characterization have established that a site is attractive, the views of affected communities must be incorporated. This is not a process of "educating" the community. Rather, it should be a true two-way dialog that recognizes that one outcome is that a site that is economically and geologically attractive may not be feasible from the standpoint of the community. There is always a chance that some stakeholders will blur the distinction between hazard and risk, and that others may make value judgments that appear to reduce the risk. Both parties must approach the process with mutual respect, and have a full and open exchange of data. One model for the process is the "continuous model" in which a common table of information is developed and checked, and the risk perceptions of all participants are explicitly incorporated.[70]

Gathering much of the data needed for site screening and characterization, to say nothing of beginning injection of CO_2, will require that the developer gain permission from surface owners to access the land. Moreover, access to the subsurface in order to drill and complete injection wells and to inject CO_2 will require the developer to secure the necessary property rights. At present, there is ambiguity about ownership of deep pore space and the right to inject CO_2 in many US states. These issues are discussed in Chapter 6.

Site Selection, Operation, and Post-injection

Once site screening and characterization has been completed, the project developer, in consultation with regulators and the local community, must decide whether to proceed with large-scale injection operations. Large-scale injection and the associated monitoring and remediation (if necessary) constitute "operation" in this book. The US regulatory process for permitting injection of CO_2 into the subsurface is discussed in Chapter 4. In addition, several entities, including the World Resources Institute,[71] Det Norske Veritas,[72] and the Canadian Standards Association, are developing or have developed best practice guidelines and standards for site operations, monitoring, and management of the post-injection phase. The means by which operational liabilities will and should be managed in the US is discussed in Chapter 7.

During the operational phase, initial pilot facilities can be expanded to accommodate large-scale injection or, if the pilot facilities are inadequate, new facilities can be constructed and connected to the pipeline system linking the site to CO_2 sources. It may be necessary to drill new wells to maintain injection rates. It will certainly be necessary to rework any abandoned wells that are not adequately plugged so as to ensure mechanical integrity and prevent migration of CO_2 or brine out of the storage unit. In regions of the world where oil and gas have been produced for decades, management of abandoned wells will pose a significant challenge, particularly for older wells, the locations and condition of which may be poorly documented or unknown.[73] Some injection wells will also be suspended (i.e., temporarily plugged) or permanently plugged before CO_2 injection ceases through neighboring wells to manage pressure in the storage unit.

Operations should be designed to use the geologic sequestration capacity effectively, and to minimize potential risks. The risks posed by geologic sequestration are driven by the buoyancy of the injected CO_2 and the increased pressure in the receiving formation. Buoyant CO_2 is trapped in the subsurface through both physical and chemical mechanisms. The physical mechanisms retain CO_2 in the formation as a separate phase, contained by geologic structures or sedimentary features, or trapped by capillary action in the fine pores of rock as a residual phase behind a migrating CO_2 plume.[74] Physical mechanisms are the dominant form of containment in the early years of operation, and will increase the security of storage over the operating life of the project.

Risks during the operational phase, shown in Figure 2.12, fall roughly into two categories—local risk to health, safety, or the environment, and global risk of CO_2

FIGURE 2.12 *Risks associated with geologic sequestration of CO_2 classified by scale of impact.*

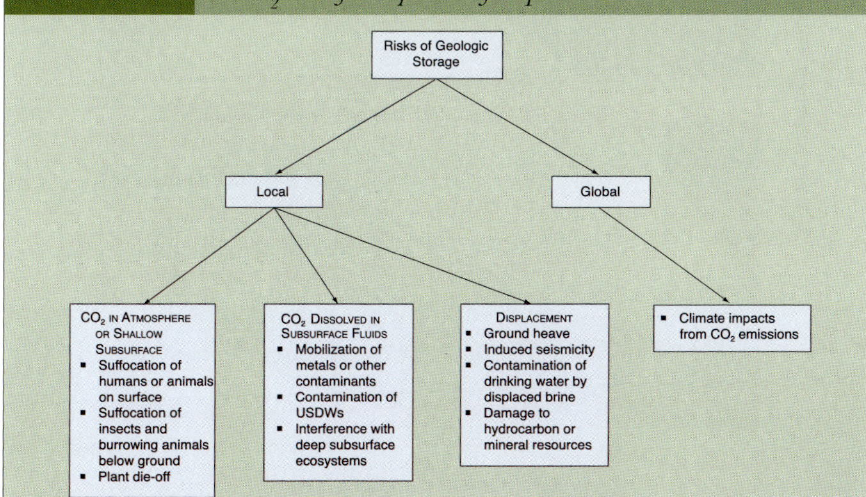

reentering the atmosphere and undermining emissions reduction goals.[75] Local impacts to drinking water could arise if CO_2 migrates out of the target formation into an underground source of drinking water or if brines displaced by injected CO_2 are forced into underground sources of drinking water. If CO_2 leaks back to the near surface or surface without impacting drinking water, it could still harm plant roots or soil ecosystems, or potentially accumulate in basements or valleys. In addition, there is the potential for induced seismicity resulting from the injection of CO_2 due to the pressure increase in the storage unit. As noted above, selection of CCS sites should involve effective and two-way dialog with the local community, with full sharing of information on the potential impacts to health, safety, and the environment and recognition by the project developers that the process may result in a decision to choose an alternative site.[76]

Over time, chemical trapping mechanisms supplement physical trapping, increasing storage security, as shown conceptually in Figure 2.13. Chemical trapping

FIGURE 2.13 *Relative contribution of trapping mechanisms to sequestration security over time. (Figure courtesy of Benson and Cook, IPCC Special Report on Carbon Dioxide Capture and Storage.)*

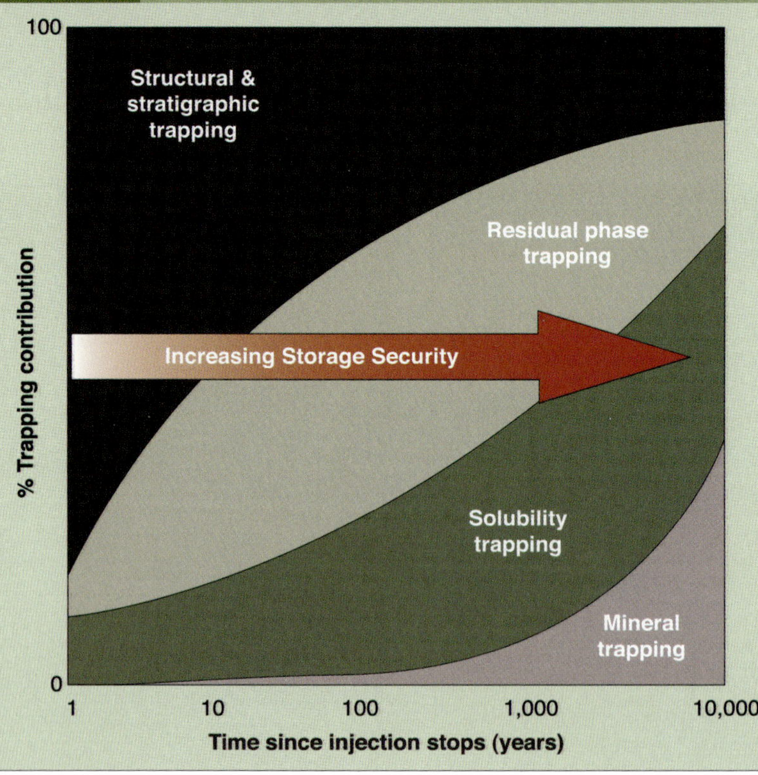

begins with dissolution of CO_2 into brine present in the formation, making the CO_2-saturated brine less buoyant than either the CO_2 phase or unsaturated brine. Dissolved carbon dioxide will react with minerals present in the formation to become immobilized—processes typically occurring on a scale of decades to centuries.[77] In addition, the pressure in the receiving formation, which reaches a peak at the end of injection, will steadily decline to a long-term equilibrium. Thus, risk associated with leakage of the sequestered CO_2, brine displacement, and induced seismicity increases through the operational period, and declines post-injection,[78] as shown conceptually in Figure 2.14.

Monitoring will be the key to safely managing risks of geologic sequestration. Commercial projects will likely measure, amongst other things, injection volumes, pressure in the injection wells and at monitoring wells, the movement of the CO_2 plume, and composition of overlying saline aquifers and underground sources of drinking water. Results from monitoring activities will serve to protect human health and the environment and ensure that CO_2 remains sequestered from the atmosphere. Monitoring will also allow geologic models to be refined, increasing confidence in the predicted behavior of the site in the long term. Box 2.3 identifies a number of the technologies available for monitoring injected CO_2 and other changes in the subsurface.

FIGURE 2.14 *Conceptual risk profile for an idealized sequestration site over time. (Adapted from Dr. Sally M. Benson, Stanford University.)*

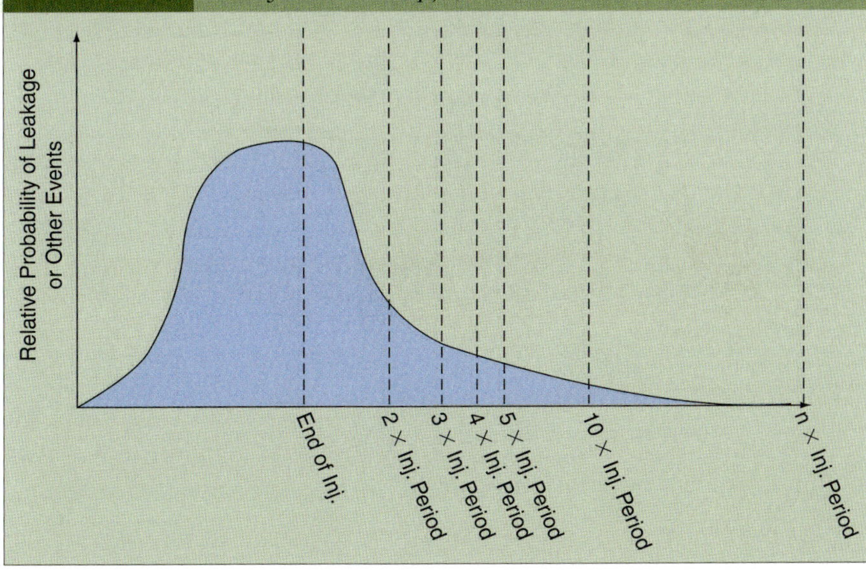

Box 2.3 *Technologies Available to Monitor the Injected CO_2 and Other Changes in the Subsurface*

Many tools have been developed for oil and gas exploration and production, natural gas storage, waste injection, and groundwater monitoring that can be applied to CO_2 sequestration.[79] However, the applicability of a monitoring tool to a CO_2 sequestration project depends on both the specifics of the site (i.e., both subsurface geology and land use at the surface) and the motivation for monitoring. The matrix below, modified from Benson et al.,[80] shows potential motivations for monitoring, available tools, and their applicability.

Motivations for monitoring at pilot scale and early projects will be different from those at commercial projects undertaken later. For example, early projects should include more intensive monitoring to increase knowledge about storage processes occurring in the subsurface and the specific location of the injected CO_2. Later, large commercial projects may focus on monitoring to verify that injected CO_2 has not migrated out of the reservoir. Tools that may be applicable to meet these monitoring objectives range from very simple and inexpensive, such as measurement of wellhead pressures and flow rates, to much more intensive and expensive, such as time-lapse 3-D seismic.

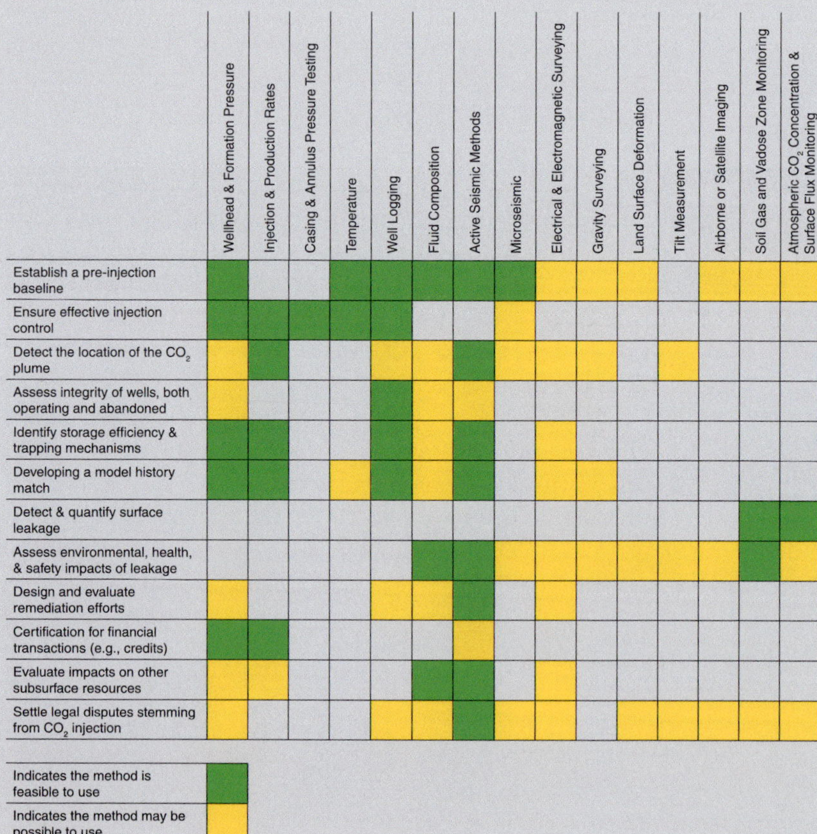

Sequestration project sites will be selected, and projects should be designed and operated, to ensure that CO_2 is retained in the subsurface. Conceptually, the amount retained can be estimated directly by measuring the mass of CO_2 in the subsurface, or indirectly by subtracting the rate of leakage from the rate of injection.

Attempts to estimate the amount of CO_2 in-place at the Sleipner project in the Norwegian North Sea, using geophysical measurements coupled with modeling, have been able to account for 50% to 80% of the CO_2 injected—the estimate varying due to assumptions about subsurface conditions and distribution of the CO_2.[81] While no leakage has been observed or is otherwise suspected from the Sleipner project, these in-situ estimates are inadequate for the purposes of greenhouse gas accounting where relatively low rates—approximately a percent of the mass sequestered per year or less—of leakage would need to be measured.[82]

In general, using geophysical methods, it is unlikely that the mass of CO_2 in-place can be estimated with greater than 20% precision; meaning retention (or leakage) cannot be estimated by measuring the mass of CO_2 in-place.[83]

Estimating retention indirectly (i.e., the amount injected minus leakage) is also hampered by a lack of precision in measurement methods. The flow rate of CO_2 injected into the formation can be gauged to a precision of approximately 2%.[84] However, it is not possible today at reasonable cost or with needed precision to measure fluxes of CO_2 between the surface of the ground and the atmosphere over hundreds of square kilometers—the typical area of a CO_2 storage project.

A number of the methods presented in Box 2.3 may be able to detect relatively low levels of leakage. If leakage is occurring from a small portion of the site, as would be expected if a leakage was occurring from a geologic feature (e.g., a fault) or well, some monitoring techniques could pinpoint and quantify this leakage, allowing emissions from the project to be appropriately accounted under climate change legislation (see Chapter 8 for further discussion).[85]

After the site reaches capacity, injection operations will cease and the "post-injection" phase will begin. In other work, this is sometimes known as the closure phase.[86] Monitoring will continue, using geophysical techniques and some monitoring wells, to document the gradual drop in subsurface pressure as the system returns to equilibrium and, potentially, to identify migration of CO_2 out of the target formation.[87] During this time, wells not needed for monitoring will be plugged. The purpose of the post-injection phase is to gain further insight about the behavior of the site and confidence in our ability to accurately predict the fate of the injected CO_2.

Long-term Stewardship

Once injection of CO_2 has stopped at a sequestration site and the site has been appropriately closed (following a post-injection monitoring period), some entity will be responsible for oversight of the facility on a permanent basis. This oversight is referred to as "long-term stewardship," and is discussed in detail in Chapter 7.

Box 2.4 *What About Safety?*

Geologists and engineers who have studied CCS at length are confident that a well-designed and properly operated project will retain CO_2 deep underground with virtually no leakage to the surface. As noted previously, the IPCC has said that the fraction of injected CO_2 retained by sequestration projects "is very likely" to be "more than 99% over the first 1000 years." The longer the CO_2 is in place, the more likely that geophysical and geochemical processes will have immobilized it. Thus, unlike some other kinds of waste for which the risk stays constant or grows over time, the risks presented by CCS can be expected to decline with time.

However, people inevitably ask, what if the experts are wrong?

There are several sources of evidence that suggest that slow leakage to the surface would pose little or no risk to public health. For example, across much of Italy, natural CO_2 of volcanic origin seeps steadily to the surface, as shown in the figure.

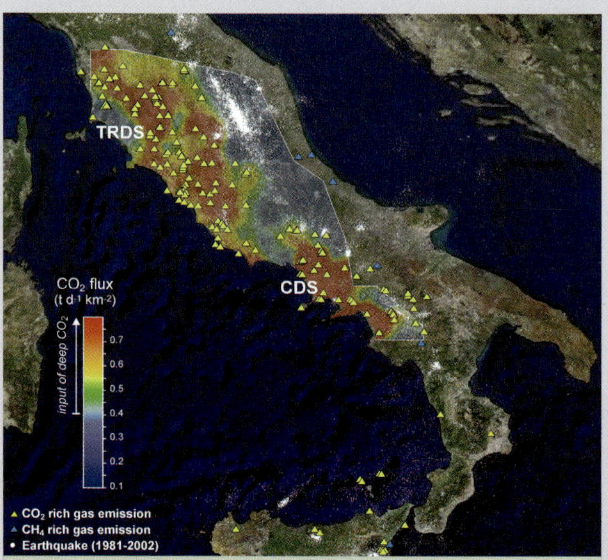

Levels of natural CO_2 seepage of volcanic origin across Italy. Yellow triangles show locations of CO_2-rich gas emissions. (Figure courtesy of Giovanni Chiodni.)

While in a few locations people have CO_2 monitors in closed spaces, such as basements, risks to public health and safety have been extremely low and easily managed. There have been a few places around the world where large natural CO_2 seeps have resulted in serious problems.[88] Thus, for example, regulators may find it prudent to monitor the levels of CO_2 if a deep, permanently stratified lake overlies an injection site.[89] Large persistent natural seepage has caused ecological damage in some locations.[90] Before such levels were reached from leakage from a CCS injection field, a variety of other indicators should make it possible to detect the problem and take corrective action.

What about the risks posed by a short-duration, large volume leak of CO_2 to the surface? Clearly, a large leak would pose some risk to the climate (the alternative pursued today, however, is 100% leakage up the stacks of power plants and other industrial facilities). Public health risk would again likely be minimal. A measurement program was undertaken by Lawrence Livermore National Laboratory of a geyser driven by CO_2 erupting from an 800 meter deep well drilled at Crystal Geyser, Utah, in 1936 as shown below. Near the well, the concentration of CO_2 was far below levels causing harm and, 100 meters from the well, modeling suggested the concentration of CO_2 was only 100 ppm higher than the background CO_2 concentration.[91] Note the people standing safely around the erupting geyser.

Measurements made by Lawrence Livermore National Labs at Crystal Geyser, Utah, suggest that the risks posed by short-duration, large volume leaks are relatively low. (Photos courtesy of Julio Friedman.)

Projects should not move into the long-term stewardship phase until modeling and monitoring demonstrates that the injected CO_2 is behaving as expected, and the projected behavior of the site indicates that there is no significant future risk of CO_2 migrating out of the storage unit or of other adverse impacts (e.g., contamination of underground sources of drinking water, etc.). At this point, open wells would likely be plugged and both the frequency and the intensity of monitoring and risk assessment activities could decrease (i.e., monitoring only if the need arises). Storage security will increase during the long-term stewardship phase, but nevertheless, some entity should be delegated to handle any problems that should arise.

When several injection projects operate in one sedimentary basin (i.e., a depression in the earth's crust in which sediments have accumulated, such as the Permian or Michigan Basins), long-term stewardship for a closed site may have to account for possible interactions and influences with adjoining active injection sites.[92]

2.7 PRACTICAL EXPERIENCE WITH GS TECHNOLOGY

As of early 2011, there are nearly 150 geologic sequestration projects operating or in various stages of development.[93] The majority of these projects are relatively small-scale pilot projects (i.e., less than a million metric tons CO_2 per year) or their planned size has not been announced. However, there is a growing number of large-scale commercial projects that are currently injecting or plan to inject 1 Mt CO_2 per year or more. The status of selected large-scale projects is summarized in Table 2.4.

In addition to the projects listed in Table 2.4, there are a large number of operating enhanced oil recovery (EOR) projects injecting natural CO_2 primarily for oil recovery, which is addressed in Section 2.8.

Processes analogous to those used in deep saline formation storage are used for acid gas disposal in Canada.[94] Through 2005, over 4.5 Mt of acid gas have been injected into deep saline aquifers and depleted hydrocarbon reservoirs at forty-eight sites, produced by forty-two separate gas plants.[95] Both H_2S and CO_2 are disposed of simultaneously because co-capture of the two gases is less expensive than separation of H_2S and CO_2. Of the total amount of acid gas injected, approximately 2.5 Mt is CO_2.[96]

Almost all of these projects are being studied to better understand the subsurface processes occurring in sequestration, means of monitoring injected CO_2 and detecting leakage, and security of CO_2 sequestration in the long term. As discussed in Chapter 5, findings from these and other pilot projects can inform development of regulations for geologic sequestration.

2.8 ENHANCED OIL RECOVERY AND ITS RELATIONSHIP TO GEOLOGIC SEQUESTRATION

Injection of CO_2 to improve recovery of oil from a reservoir is known as CO_2-flood enhanced oil recovery (CO_2-EOR). CO_2-EOR has been a commercial activity in the US since the 1970s. Today in the US, approximately 270,000 barrels of oil per day (BOPD) are produced from over 100 projects.[97] There are also approximately a dozen CO_2-EOR projects operating in Brazil, Canada, Trinidad, and Turkey. Thus, there is considerable experience with development of CO_2-EOR projects in the US and elsewhere. The principal motivation to develop CO_2-EOR projects has been to increase the amount of oil that can be extracted from aging oil fields and, particularly in the US, slow the rate of decline of domestic oil production.

The CO_2-EOR process can be broken down into several steps: relatively pure CO_2 is injected through an injection well into the target reservoir; the CO_2 extracts oil from the reservoir; and, the CO_2-oil mixture, along with reservoir brine, is brought to the surface at a production well. These three process steps happen at the level of a "well pattern," a cluster of injection and production wells. Depending

Table 2.4 *Selected planned and operating large geologic sequestration projects as of early 2011.*

Project	Location	Operator	Storage Type	Injection Start Date	Sequestration Rate	Total Planned
Sleipner	North Sea	Statoil	Aquifer	1996	1 Mt/y	20 Mt
Weyburn-Midale	Saskatchewan, Canada	EnCana	EOR	2000	3 Mt/y[a]	19 Mt
In Salah	Sahara, Algeria	Sonatrach, BP, Statoil	Depleted Gas Reservoir	2004	1 Mt/y	17 Mt
Snohvit	Melkøya, Norway	Statoil	Aquifer	2007	0.7 Mt/y	
Gorgon	Barrow Island, Australia	Chevron	Aquifer	2015[b]	3.4 Mt/y	>100 Mt

Notes: a. The average annual sequestration rate over the life of the project will be approximately half.
b. Planned date.

on the size of the oil field, there may be hundreds of patterns. For example, at the Wasson Denver Unit in Texas, which is among the largest CO_2-EOR projects (by production), there are approximately 1,000 injection wells and 600 production wells spread across 28,000 acres (44 square miles).[98] At the surface, the produced fluid mixture—oil, CO_2, light hydrocarbons, and typically H_2S—from a number of well patterns is collected and separated by phase. Hydrocarbons are treated and transported for sale, and CO_2, separated from the produced oil, is compressed and recycled for injection along with new CO_2.

Interest in developing CO_2-EOR projects increases with the price of oil, as a large project is costly and requires sustained high oil prices to be commercially viable. Currently, production from operating CO_2-EOR projects and development of new projects in the US is limited by the cost and availability of CO_2.[99] Therefore, traditional CO_2-EOR projects are designed to minimize the amount of CO_2 that must be purchased by recovering and recycling as much produced CO_2 as possible. CO_2-EOR projects explicitly intended to be operated for geologic sequestration could be designed somewhat differently from traditional CO_2-EOR projects. For example, it may be possible to optimize operation and design of injection patterns to maximize both oil recovery and CO_2 storage.[100] It could also be possible to eliminate CO_2 recycling (and potentially H_2S separation) from the CO_2-EOR operation and instead sequester the produced CO_2 in a deep saline formation, maximizing the amount of CO_2 sequestered while recovering just as much oil as from a traditional EOR project.

Typically between 0.16 and 0.42 metric tons of CO_2 are purchased to produce one barrel of oil,[101] and upwards of 50% of the CO_2 injected is recycled. Thus, between 0.32 and 0.85 tonnes of CO_2 are injected to recover one barrel of oil. Using these estimates for CO_2 utilization, hundreds of millions of tonnes of CO_2 have been purchased for CO_2-EOR since commercial operations began in the US to date; approximately 80% of this CO_2 has come from natural accumulations of CO_2, such as McElmo Dome, located in southern Colorado, and Jackson Dome in Mississippi, with lesser amounts from relatively pure anthropogenic sources of CO_2 such as gas processing and fertilizer plants.[102]

Many studies have highlighted the potential economic and energy security benefits of applying CO_2-EOR in the US[103] and elsewhere.[104] One recent study estimates that between 70 billion and 110 billion additional barrels of oil are technically recoverable from US oil reservoirs in the lower-48 states through CO_2-EOR depending on the design and operation of projects; of this total, between 40 billion and 60 billion barrels are economically recoverable at an oil price of $70 per barrel and a CO_2 supply price of $15 per tonne, resulting in sequestration of approximately 10 Gt of CO_2.[105] For comparison, the US Energy Information Administration (EIA) reports proven (1P) crude oil reserves in the lower-48 states to be approximately 17 billion barrels at the end of 2009. While the addition of 50 billion barrels to proven reserves through use of EOR is unlikely given historical reserve growth rates,[106] it is nonetheless clear that widespread application of CO_2-EOR could result in large additions to domestic oil resources.

The benefits of CO_2-EOR as a means of geologic sequestration are less clear-cut. Increasing the use of CO_2-EOR offers three benefits: (1) expanding the knowledge base needed for commercial scale CCS; (2) encouraging construction of infrastructure, such as pipelines, that would benefit wider deployment of CCS; and (3) sequestering CO_2. As discussed in the following paragraphs, current economic and regulatory drivers fail, by and large, to drive CO_2-EOR to generate these benefits.

Technical knowledge transfer from CO_2-EOR to geologic sequestration in saline formations is limited by substantive differences between the processes, and by commercial considerations that impede transparency. For example, understanding of reservoir pressure effects is a key knowledge gap for GS in saline aquifers, but large-scale pressure effects are not observed in CO_2-EOR because reservoir pressure is managed by the production of fluids in direct proportion to the volume of CO_2 injected. Likewise, more experience with methods to measure CO_2 retention and monitor for leakage are needed to support commercial-scale CCS, but in general, monitoring at CO_2-EOR projects is not designed with these objectives in mind. The current regulations for CO_2-EOR do not require monitoring for environmental impacts, and these activities would only increase the cost of producing oil.

The sequestration value of CO_2-EOR is less certain than that of sequestration in saline aquifers both because CO_2-EOR projects are, at the current time, not monitored, as discussed above, and more importantly, because the process enables the recovery of additional fossil fuel. Life-cycle assessment of CO_2-EOR projects that use CO_2 captured from coal-fired power generation have shown that these projects result in additional CO_2 emissions above those that would be expected from a similar project sequestering CO_2 in a deep saline aquifer due to the combustion of the produced oil.[107] However, the project may still result in an emissions reduction if there is displacement of energy (in the form of electricity and oil) that would otherwise come from high carbon intensity sources (e.g., traditional pulverized coal and oil sands, respectively).[108] Similarly, a CO_2-EOR project using CO_2 captured from power generation could result in an increase in emissions if the energy produced from the project displaces energy from low carbon intensity sources (e.g., light, sweet crude oil, and wind energy). The challenge is that displacement requires that a barrel of oil produced from CO_2-EOR is not produced elsewhere at any time in the future. Thus, geologic sequestration (via CO_2-EOR or otherwise) must be implemented as part of a broader, cross-sectoral emissions reductions program. If sequestration via CO_2-EOR is developed in the absence of such a policy, life-cycle methods should be applied to estimate net emissions from the project.

One final difference between CO_2-EOR and GS in saline formations is the adequacy of the supporting legal and regulatory framework. Current federal and state laws and rules regarding CO_2 injection facilitate CO_2-EOR and, in general, the industry is content with the status quo.[109] The challenge is to craft a legal and regulatory framework that can equally support deployment of large-scale commercial CCS in saline aquifers solely for the purpose of emissions reductions. This book does not focus on legal and regulatory barriers to widespread use of CO_2-EOR, but instead on those faced by projects sequestering CO_2 principally for the purpose of emissions reductions. In some cases, addressing the barriers facing sequestration in saline aquifers will also enable greater deployment of CO_2-EOR.

3

SITING CO_2 PIPELINES FOR GEOLOGIC SEQUESTRATION

If CCS is to become a viable tool for greenhouse gas (GHG) reductions, significant investments in infrastructure will be necessary, including in the development of a CO_2 pipeline network. The ability to transport very large volumes of CO_2 via pipeline from source to sequestration sites will be crucial to the deployment of large-scale CCS projects. In the US today, there are both point-to-point CO_2 pipelines, delivering CO_2 from a single CO_2 source to a single location, as well as CO_2 pipeline networks that collect CO_2 from one or more sources and deliver it to multiple locations spread across several states. These pipelines were built to support CO_2-flood enhanced oil recovery (EOR) operations. They carry roughly 60 Mt of CO_2 per year over 5,800 km of pipeline.[1]

Figure 3.1 shows the locations of the existing US CO_2 pipeline infrastructure. While there is considerable uncertainty over the capacity, length, and geographic distribution of the pipeline infrastructure required for large-scale CCS (see Chapter 2), studies have suggested that the existing infrastructure eventually may have to be expanded tenfold to accommodate CO_2 transport for CCS.[2] It is also clear that lower costs and higher reliability can be obtained from interconnected CO_2 pipelines (i.e., networks) rather than a point-to-point infrastructure.[3] This build out, and the emergence of larger CO_2 pipeline networks, will occur under a different business model than that which supported the existing CO_2-EOR pipelines.

We believe that in addition to capital, system expansion will require substantial regulatory reforms. The regime in place today, developed for the EOR community, is based on regulation by individual states. To build the interstate system that will be necessary to support large-scale CCS, the option for regulation at the federal level will be needed to avoid the inefficiencies inherent in a system of state-by-state oversight. Such federal regulatory reform will need to address federal siting process, including eminent domain authority to site pipeline right-of-ways similar to the

FIGURE 3.1 *The existing US CO_2 pipeline infrastructure.*

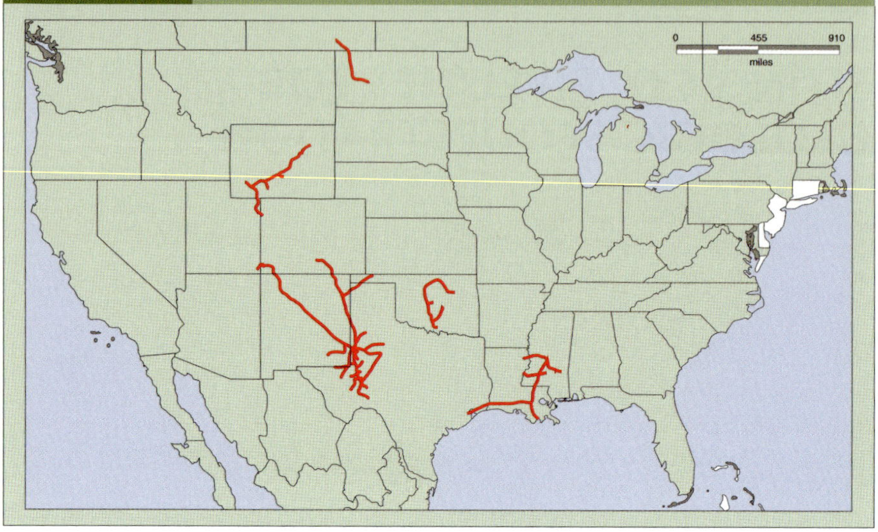

scheme for interstate natural gas pipelines. It will also need to provide a streamlined federal permitting process if the project is on or crosses federal lands. In our view, without a consistent federal regulatory system, the infrastructure growth necessary for wide spread deployment of CCS will be hampered.

To understand the role played by infrastructure, this chapter summarizes the current state of CO_2 pipeline regulation under federal and state law, evaluates existing law in areas that may be important for a national CO_2 pipeline system, discusses alternative regulatory frameworks that could be considered to support development and operation of a much larger CO_2 pipeline network, and concludes with a recommendation for reform. Chapter 2 provides background information about the economics of CO_2 transport by pipeline and reviews the potential future scale of an interstate pipeline system in the US.

3.1 EXISTING FEDERAL REGULATION OF CO_2 PIPELINES

Federal law governing CO_2 pipelines is inconsistent, confused, and uncertain. Today, there is no federal scheme for siting CO_2 pipelines, and a minimal scheme for regulating rates and access that is limited to infrastructure sited on federal lands. The two agencies with the most regulatory expertise and closest form of a statutory mandate to regulate CO_2 pipelines, the Federal Energy Regulatory Commission (FERC) and the Surface Transportation Board (STB), at present do not exercise jurisdiction over CO_2 pipelines.

FERC possesses jurisdiction to regulate transportation and sale at wholesale of natural gas in interstate commerce under the Natural Gas Act (NGA).[4] A pipeline operator cannot engage in the transportation or sale of natural gas subject to FERC jurisdiction or construct, extend, or acquire a natural gas pipeline without obtaining a certificate of public convenience and necessity from FERC for these purposes.[5] The Commission will issue a certificate only if "required by the present or future public convenience and necessity."[6] FERC can impose conditions on the certificate[7] and has the power to determine the service area to be covered.[8] Perhaps the most valuable tool in the NGA is the right of eminent domain granted to the holder of a certificate of public convenience and necessity.[9] These provisions from Section 7 of the NGA, combined with Section 4 (rates and charges) and Section 5 (fixing rates and charges), have led the courts to repeatedly interpret the NGA as providing for exclusive and preemptive federal jurisdiction over siting of interstate natural gas pipelines.[10] However, FERC has disclaimed jurisdiction to regulate CO_2 based on a finding[11] that CO_2 is not a "natural gas" under the Natural Gas Act.[12]

FERC also regulates oil pipelines under the Interstate Commerce Act (ICA).[13] The Commission's responsibilities include: 1) regulation of rates and practices of pipeline companies engaged in interstate transportation of petroleum and petroleum products; and 2) establishment of non-discriminatory conditions of service in order to provide shippers access to pipeline transportation. However, there has been no discussion to fold regulation of CO_2 pipelines into the ICA. In fact, as will be discussed later on, quite the opposite has occurred.

Meanwhile, the STB, an independent federal administrative agency within the Department of Transportation, has neither disclaimed jurisdiction in the same manner as FERC, nor asserted jurisdiction over such pipelines. The STB is responsible for economic regulation of certain common carrier interstate transportation. This responsibility primarily relates to railroad transportation, but also includes interstate transportation by pipeline of commodities "when transporting a commodity other than water, gas or oil," with the term "gas" undefined.[14] However, the STB's predecessor agency, the Interstate Commerce Commission (ICC), found that CO_2 is a "gas" and therefore non-jurisdictional under the ICA when transported by pipeline.[15] Should STB follow this precedent, it would not regulate CO_2 pipelines either.

To date, the only federal agency that has exercised any sort of authority over CO_2 pipelines' rates or access is the Bureau of Land Management (BLM). BLM is one of the federal agencies with authority to grant rights-of-way (ROW) across federal lands and has the responsibility for administering ROW on federal lands managed by the Department of the Interior. BLM imposes the equivalent of a common carrier obligation on CO_2 pipelines crossing federal lands on the basis that CO_2 is a "natural gas" within the meaning of the Mineral Leasing Act (MLA),[16] one of that agency's governing statutes. The MLA provides that:

> Rights-of-way through any Federal lands may be granted by the Secretary of the Interior or appropriate agency head for pipeline purposes for the

> transportation of oil, natural gas, synthetic liquid or gaseous fuels, or any refined product produced therefrom.[17]

The common carrier obligation arises once a ROW is granted under the MLA, at which point BLM imposes an obligation on the pipeline to "accept, convey, transport, or purchase without discrimination all oil or gas delivered to the pipeline without regard to whether such oil and gas was produced on Federal or non-Federal lands."[18]

The conflicts inherent in the definitions of CO_2 are palpable. Whereas FERC cannot regulate CO_2 because it is not a "natural gas" under the Natural Gas Act, the BLM has authority over CO_2 pipelines because CO_2 has been found to be a "natural gas" under the MLA. In the meantime, the STB would be prohibited from regulating CO_2 pipelines at all under the ICA, if CO_2 is defined simply as a "gas." This confused state of affairs arises from a series of federal court and agency decisions that pre-date any concepts of CCS as a GHG mitigation tool by many years.

In *Cortez Pipeline Co.* (*Cortez*), FERC found that it did not have jurisdiction over CO_2 pipelines under the NGA, even when they transport small amounts of natural gas.[19] Cortez sought to develop a pipeline connecting a CO_2 reservoir in Colorado with oil fields in Texas for EOR. Cortez requested that FERC issue a declaratory order stating that FERC did not have jurisdiction over the proposed pipeline because the supercritical fluid being transported was not "natural gas" within the meaning of the NGA. (The NGA defines natural gas as "natural gas unmixed or any mixture of natural and artificial gases."[20]) The pipeline company stated that the mixture transported in the pipeline project would be 98% CO_2, with the other 2% of mixed composition, including methane.[21]

In response to the request, FERC analyzed the NGA to determine whether the CO_2 and methane gas mixture was "natural gas" within the meaning of the statute. FERC looked beyond a scientific or technical definition of "natural gas" to determine its jurisdiction, and instead looked to the reasons for the passage of the NGA.[22] FERC noted a lack of debate over any ambiguity in the term "natural gas" during NGA enactment. It determined that the only debate in the legislative history around the term "natural gas" in the NGA focused on "whether unmixed artificial gas should be included" in the definition, concluding that, "[i]t seems likely that Congress used the common meaning of 'natural gas' of a mixture of gases, including a sufficient component of hydrocarbons to give it heating value."[23]

After FERC determined that there was no specific chemical composition under the NGA that constitutes "natural gas," FERC evaluated Congress's objectives in enacting the NGA. FERC stated that the "goal of the NGA was to protect the consumers of a salable commodity from 'exploitation at the hands of the natural gas companies' and was framed to afford consumers a bond of protection from excessive rates and charges."[24] It considered whether to include the CO_2 pipeline within its jurisdiction "in light of the general goal of the NGA," and found that "no goal or purpose of the NGA" would be advanced by FERC jurisdiction over the CO_2 pipeline,[25] and accordingly did not assert jurisdiction.

In 1981, the ICC determined that it lacked jurisdiction over CO_2 pipelines in a proceeding involving the same pipeline project as the FERC decision in *Cortez Pipeline Co.* As it did before FERC, Cortez filed a petition with the ICC for a declaratory order that CO_2 pipeline transport is exempt from ICC jurisdiction.[26] Cortez argued that the Interstate Commerce Act (ICA) specifically excluded from ICC jurisdiction interstate pipeline transportation of "water, gas, or oil," and that CO_2, while transported as a supercritical fluid, is a gas at atmospheric pressure, the transportation of which falls within the statutory exemption from regulation.

The ICC proceeded to analyze the situation in terms of the meaning of "gas" in the statutory exemption beginning with the history of the Hepburn Act of 1906, which granted jurisdiction over common carrier pipelines to the ICC.[27] The ICC found that the original language in the Hepburn Act provided ICC jurisdiction over interstate commodity transportation "except water and except natural or artificial gas."[28] The term "natural or artificial" was eliminated in a 1978 ICA recodification, but the ICC determined that the term's removal was not a substantive change.

Based on its reading of the Hepburn Act and the legislative history, the ICC issued a preliminary finding that it lacked jurisdiction over CO_2 pipelines because "the universe of gas types classified by origin or source was excluded."[29] The ICC explained that the decision of FERC, as a "sister agency, should be given weight if possible."[30] However, the ICC distinguished FERC's decision, because it was not based on an interpretation of the term "natural and artificial gas." After receiving only supportive public comments on its preliminary decision, the ICC affirmed the preliminary decision that it did not possess jurisdiction over CO_2 pipelines.[31] The ICC found that based on the plain meaning of the statutory exemption for "water, gas or oil,"[32] and the legislative history of the Hepburn Act of 1906, "all gas types classified by origin or source are excluded from our jurisdiction. Consequently, carbon dioxide gas, the subject of the petitions, is also excluded, when transported by pipeline."[33]

The ICC precedent has been questioned after passage of the ICC Termination Act of 1995 (ICCTA),[34] which abolished the ICC and created the STB to serve as the ICC's successor agency. In 1998, the General Accounting Office (GAO) released a report that specifically found that CO_2 pipelines are within the oversight authority of the STB, along with at least one other gas, hydrogen.[35] To date, the STB has not heard any case specifically requesting it to rule on its jurisdiction over CO_2 pipelines, and on that basis has declined to address the jurisdictional issue raised in the GAO report.[36]

While the STB is not bound by the ICC ruling,[37] the statutory language that was interpreted in the ICC's *Cortez* decision is virtually identical to that in the corresponding section of the current ICCTA.[38] Given the ICC's status as a predecessor agency and the similarity in statutory language, if the question arises, STB may be inclined to follow the ICC *Cortez* decision with respect to jurisdiction over CO_2 pipelines. Whether such a decision could be sustained on judicial review remains to be seen. The ICC's review of the legislative history of the 1906 Hepburn Act, in the underlying decision, fails to support its conclusion that all gases, rather than combustible gases, were intended to be covered.

The BLM's decision to treat CO_2 pipelines as common carriers derives from *Exxon Corp. v. Lujan.*[39] Exxon challenged the issuance of a ROW across federal lands under the MLA, instead arguing that the ROW should have been issued under the Federal Land Policy and Management Act (FLPMA).[40] FLPMA provides that the Secretary shall issue ROW for:

> pipelines and other systems for the transportation or distribution of liquids and gases, *other than water and other than oil, natural gas, synthetic liquid or gaseous fuels, or any refined product produced therefrom*, and for storage and terminal facilities in connection therewith.[41]

A FLPMA ROW, unlike a MLA ROW, does not require that the operator act as a common carrier.[42] Exxon challenged the ROW at the agency and district court level, arguing that CO_2 is not a "natural gas" within the meaning of the MLA, but rather falls within the purview of FLPMA.

The US Court of Appeals for the Tenth Circuit (Tenth Circuit) addressed this statutory interpretation question by reference to common administrative law principles under which courts defer to an agency's interpretation of a statute it implements where the statute is ambiguous and the agency interpretation is reasonable.[43] The Tenth Circuit noted that there are varying definitions of natural gas and that courts and agencies have interpreted the meaning of natural gas in different contexts, including in the FERC *Cortez* case. The court concluded that the statutory term "natural gas" was ambiguous. It looked further to the legislative history of the MLA, but concluded that "the legislative history of the MLA does not establish Congress's intention with the requisite clarity."[44] Consequently, the court applied the *Chevron* doctrine and granted deference to the BLM's reasonable interpretation of the MLA to cover ROW for CO_2 pipelines, based on an interpretation that CO_2 was "natural gas" under the MLA.

Because each agency's jurisdiction is a creature of statute, these dueling interpretations and definition of CO_2 are unlikely to change without some new statutory authority provided by Congress. However, even if the existing statutes were modified to explicitly permit regulation of CO_2 pipelines, some agencies may be better suited for the task than others. For example, even if one assumes that the STB has jurisdiction over CO_2 pipelines, the STB's regulatory oversight would be limited. Importantly, the STB authority, unlike FERC authority under the NGA, does not encompass siting, certification, or eminent domain authority with respect to pipelines it regulates. Instead, the STB's regulatory role is to ensure that a common carrier pipeline: 1) charges reasonable, non-discriminatory rates;[45] 2) establishes classifications, rules, and practices on matters related to its transportation and service;[46] 3) does not subject its shippers to unreasonable discrimination;[47] 4) provides proper facilities for the interchange of traffic;[48] and 5) provides transportation and service, as well as rates and other terms of service, upon reasonable request.[49]

Even if the STB exercised regulatory authority over CO_2 pipelines, its jurisdiction over a particular pipeline would depend upon whether the pipeline company is a "pipeline carrier."[50] The ICCTA defines "pipeline carrier" as a "person providing pipeline transportation for compensation."[51] If the company entered into transactions with other companies to ship their carbon dioxide in interstate commerce, then the company would be a "pipeline carrier" and subject to STB regulation (assuming, again, that supercritical CO_2 is not an exempt gas under the ICA). In addition, according to the precedent established pursuant to the ICA, a pipeline that does not engage in "transportation" is not subject to regulation. For example, if a company owned or operated pipelines in which it shipped only CO_2 it had produced, it would not be engaged in interstate "transportation" within the meaning of Title 49.[52] This precedent is consistent with the ICCTA definition of a "pipeline carrier," and would seem to indicate that if a CO_2 capturer owns the pipelines that transport only CO_2 it produces from its own facilities, it would not be regulated under Title 49. In the EOR community, many of the CO_2 pipelines would not be considered "pipeline carriers" or engaged in interstate "transportation" because of these reasons even if CO_2 pipelines were subject to STB regulation.[53] However, it is not yet apparent whether CCS pipelines would have similar business models. Although it is unlikely that each emission source would have a dedicated pipeline directly to a sequestration site due to issues of cost and scale.

FERC, on the other hand, has far more rigorous regulatory authority over natural gas pipelines under the NGA. It may begin an investigation into a pipeline's rates on its own initiative while the STB may begin investigations only in response to complaints by shippers or other affected parties.[54] In addition, the STB has no authority to require a pipeline carrier to file their rates or regulate a pipeline carrier's decision to enter or abandon markets.[55] The NGA provides FERC with clear authority to require natural gas pipelines to do so. The BLM, which is a federal land management organization as opposed to a pipeline regulator, would be an illogical and ineffective agency to assign comprehensive CO_2 pipeline regulation authority.

Lack of a Federal Eminent Domain Authority for CO_2 Pipelines

Under Section 7 of the NGA, FERC may grant eminent domain authority to natural gas pipeline developers when the proposed project is found to be a matter of public convenience and necessity.[56] This provision has encouraged the successful build-out of a large interstate natural gas pipeline network. By contrast, no comparable federal eminent domain authority exists for siting CO_2 pipelines. Lack of federal eminent domain authority is not fatal to a project if the states have a useful corollary to site projects. A handful of states with active EOR business communities provide CO_2 pipelines with eminent domain authority.

However, even if a state has a scheme in place, it may not be a sufficient substitute for federal siting and eminent domain authority. For example, New Mexico, as explained below, authorizes eminent domain authority if the CO_2 pipeline is for a "public use." In an interstate pipeline context, where a pipeline may be passing through a state without providing any direct benefit to the residents of that state, "public use" in that particular state may be an open question and a grant of eminent domain authority less certain. This issue has plagued the siting of high voltage interstate electric transmission projects, which are seen by some as necessary to interconnect distant renewable energy sources like wind and solar. State commissions may show reluctance to certify certain interstate projects that merely pass through a state onto larger markets because their consumers are not the projects' direct beneficiaries.[57]

Similarly, CCS may require the transportation of CO_2 over many hundreds of miles of pipeline from emission site to a suitable sequestration site. At best, this could lead to a time-consuming state-commission-by-state-commission approval process. At worst, state opposition may block a project if it perceives the pipeline as providing no direct consumer benefit.

The Energy Policy Act of 2005 (EPAct 2005) attempted to remedy the interstate transmission siting problems by enacting section 216 of the Federal Power Act (FPA). Section 216 provides "back-up" federal transmission siting authority in certain limited circumstances. It authorizes FERC to issue permits for the siting, construction, or modification of transmission facilities in constrained areas designated by the Secretary of Energy as national interest energy transmission corridors when a state commission has withheld approval of an application for more than one year or imposed conditions that make the project economically infeasible.[58] This limited grant of new federal authority has been controversial, and to date, no project developer has been able to utilize it. Also, in 2009, the US Court of Appeals for the Fourth Circuit overruled FERC's interpretation that "withholding approval" of an application included denial of the application.[59] Thus, it has yet to have a direct impact on interstate transmission siting or serve as a basis for federal eminent domain. While section 216 might have served as a model for a future CO_2 pipeline eminent domain scheme prior to the Fourth Circuit's ruling, its authority is now too ambiguous. However, section 216's back-up federal siting model, if redrafted, could serve as a model.

According to some, the debate over state- or federal-level siting of CO_2 pipelines is premature. A recent report by the Interstate Oil and Gas Compact Commission (IOGCC) determined that the state-by-state regulatory process was not a hindrance to the market-driven development of CO_2 pipelines for EOR development.[60] However, the report acknowledges that market forces will not be the sole driving factor for CCS pipelines. Also, for large-scale CCS to be viable, the pipeline network will need to move CO_2 from regions with limited sequestration capacity due to poor geology (e.g., North Carolina[61]) or land-use constraints, to

FIGURE 3.2 *Power plants located in regions that do not have saline formations suitable for sequestration (yellow ovals) would have to transport their CO_2 via pipeline to regions that do. Similarly, plants located in regions with thin or low-capacity reservoirs (such as western Pennsylvania) might find it more cost effective to ship their CO_2 via pipeline to regions that have large-capacity reservoirs (such as Ohio). Regions shaded in blue show assessed saline aquifers. The map of saline aquifer locations is from NETL.*

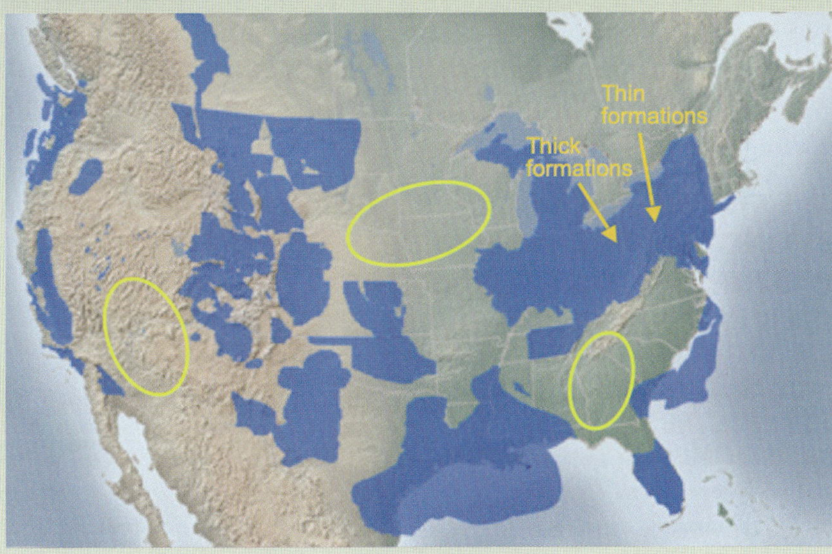

regions with suitable sequestration sites.[62] Moreover, even when a CO_2 source is located in proximity to suitable sequestration sites, savings could be realized by shipping CO_2 over significant distances to larger, lower-cost sequestration projects (Figure 3.2).[63] This will require interstate transportation well beyond what is currently needed for the EOR community.

Federal Safety Regulation of CO_2 Pipelines

Safety regulation of CO_2 pipelines is clearly established and does not suffer from the same uncertainties as siting and economic regulation of those pipelines. The US Department of Transportation's (DOT) Pipeline and Hazardous Materials Safety Administration (PHMSA) has primary authority to regulate CO_2 transportation pipelines under the Hazardous Liquid Pipeline Safety Act (HLPSA) of 1979.[64] The HLPSA directs PHMSA to regulate CO_2 transported by a hazardous liquid pipeline facility.[65] As a result, PHMSA currently regulates CO_2 transported in the liquid or supercritical state, but not as a gas.[66] Within PHMSA, the Office of Pipeline Safety (OPS) regulates the design, construction, operation, maintenance, and spill response planning for regulated hazardous liquid pipelines, including CO_2 pipelines.[67] PHMSA establishes minimum safety standards for CO_2 pipelines, and has largely preempted states from establishing their own standards for interstate pipelines.[68] With respect to safety, pipelines transporting CO_2 are regulated to the same degree as those transporting hazardous liquids.[69]

While the above scheme is well defined, DOT is aware that it may need to change its regulations to accommodate CCS. It recently issued a proposal to reauthorize the Pipeline Safety Laws that would amend § 60102(i) to include CO_2 transported as a gas.[70] The proposed language reflects DOT's belief that "carbon sequestration projects may involve pipelines transporting carbon dioxide as a gas."[71]

3.2 EXISTING REGULATION OF SITING, RATE SETTING, SAFETY, AND ACCESS TO CO_2 PIPELINES IN SELECTED STATES (TEXAS, NEW MEXICO, OHIO, AND PENNSYLVANIA)

The lack of clear federal regulation over CO_2 pipelines' siting, rates, and access does not mean they go unregulated. As a general matter, the states and not the federal government are responsible for siting both interstate and intrastate CO_2 pipelines that do not cross federal lands, and for regulating their rates and access (i.e., common carrier or contracted capacity). Because the largest CO_2 pipeline in existence today, operated by Kinder Morgan CO_2, is sited primarily in Texas and New Mexico, the authors undertook a case study to understand the regulatory schemes for siting, access, and rates in both of those states. These case studies determined that Texas and New Mexico have mature regulations governing CO_2 pipelines because EOR production wells have existed there for decades. However, case studies were also conducted on Pennsylvania and Ohio, states without EOR production, but where future CO_2 pipelines would be necessary to transport emissions from that region's coal-fired power plants.

Texas

The Texas Railroad Commission (TRC) regulates the rates and access to Texas intrastate CO_2 pipelines if they choose to become a common carrier. Notably, in Texas, CO_2 pipelines have the option to choose to become a common carrier,[72] which, in return for certain rights, imposes certain obligations on the pipeline. An intrastate CO_2 pipeline regulated as a common carrier is required to charge equal rates for like service,[73] and to "make and publish their tariffs under rules prescribed by the [Texas Railroad Commission]."[74] The Texas Railroad Commission does not appear to prescribe detailed tariff provisions for CO_2 pipelines, as it does for petroleum pipelines.[75]

Once a CO_2 pipeline becomes a common carrier in Texas, it has the statutory right of eminent domain. That is, the pipeline developers may take lands for the public use of pipeline infrastructure development. Lands for pipeline construction are often obtained through leases, with the threat of eminent domain action looming over the transactions. The Texas Natural Resources Code provides that:

> (a) Common carriers have the right and power of eminent domain.
> (b) In the exercise of the power of eminent domain granted under the provisions of Subsection (a) of this section, a common carrier may enter on and condemn the land, rights-of-way, easements, and property of any person or corporation necessary for the construction, maintenance, or operation of the common carrier pipeline.[76]

In the exercise of the power of eminent domain, property owners are entitled to just and adequate compensation for the public use of their land. The standard easement granted in Texas is 50 feet wide.[77] Of note, Texas does not require CO_2 pipeline operators to obtain a certificate of need and public convenience before the power of eminent domain is granted.[78] Siting is not performed by the state, but by the pipeline operator, which has the authority to decide the route a pipeline takes.[79] The Safety Division of the Railroad Commission of Texas oversees pipeline construction and grants permits for operations of intrastate hazardous liquids pipelines.

New Mexico

Unlike Texas, New Mexico currently does not have a regulatory scheme in place to regulate the rates of intrastate CO_2 pipelines.[80] However, New Mexico does have a scheme in place to provide for eminent domain authority, although the pipeline need not be a common carrier in order to exercise eminent domain authority.[81] The eminent domain right applies to trunk lines only, which are primary transportation lines.

New Mexico's eminent domain statute provides for the authority to condemn surface property for pipeline construction and specifically includes CO_2 pipelines. The statute allows "any person, firm, association or corporation" to obtain a "right-of-way for the construction, maintenance and operation of [such] pipelines" and to enter onto state and private lands to make necessary surveys and examinations for them.[82] The state has extensive procedural requirements in place for eminent domain proceedings.[83] Should a dispute arise over condemned property, New Mexico will allow the condemner to take possession if it can show that the property condemned is for public use.[84] The definition of "public use" in New Mexico is circular. Condemnation for the provision for CO_2 pipelines is considered "public use" based on the legislature's decision to grant such pipelines eminent domain authority.[85]

Ohio and Pennsylvania

Unlike Texas and New Mexico, CO_2 is not used for EOR in Ohio or Pennsylvania and there are no CO_2 pipelines currently sited there. Unsurprisingly, there are no state statutory or regulatory schemes that specifically support the siting of, access to, or rates for CO_2 pipelines. For example, Ohio permits companies that transport "natural or artificial gas, petroleum, coal or its derivatives . . . through . . . pipes" to obtain eminent domain authority to condemn property for pipeline construction.[86] To support the construction of CO_2 pipelines without amendment, Ohio would need to define CO_2 as an "artificial gas." Under the current scheme, artificial gas is defined as a type of fuel within a list of fuels used primarily for their energy content.[87] Carbon dioxide for CCS purposes does not meet that definition. Furthermore, the state's public utility commission has no authority over CO_2 pipelines for either siting or ratemaking purposes because the statutory definition of a public utility does not encompass an entity that transports CO_2.[88] Rather, there is no state agency authorized to regulate CO_2 pipelines. Pennsylvania does not have specific authorities on intrastate CO_2 pipeline siting or rates. It does grant eminent domain and ratemaking powers to public utility corporations, but these do not extend to CO_2 pipelines.[89]

Ohio and Pennsylvania are contemplating EOR and CCS development in their states, which could lead to further action on CO_2 pipeline siting, access, and regulation. For example, Ohio updated its oil and gas laws in June 2010 to require a permit from its Division of Mineral Resources Management in order to inject CO_2 for EOR purposes.[90] This is the first time in that state that CO_2 has been legislated in that manner. Also, state policymakers are likely to be aware of a 2008 Pew Center on Global Climate Change paper analyzing the EOR potential in Ohio, which also modeled a hypothetical 450 mile intrastate CO_2 pipeline.[91] Furthermore, the US Department of Energy's Carbon Sequestration Atlas of the US and Canada shows a significant sequestration resource that could be realized in Ohio.[92]

Pennsylvania is also contemplating CO_2 pipeline regulation. PA House Bill 2405, which was introduced on April 23, 2010, but tabled in September 2010, would provide for state regulation of CO_2 transport rates as well as leasing of state land for the purposes of constructing CO_2 pipelines for carbon sequestration. The Bill also proposes to treat transporters of CO_2 as public utilities. Although it has not passed, Pennsylvania did require its Department of Conservation and Natural Resources to perform a study on the development of a state-wide carbon dioxide sequestration network, including pipelines, by April 2009.[93] This was published in November 2009.[94] The report concluded that within Pennsylvania "there is a major risk that it would be extremely difficult, if not impossible to site the pipelines without eminent domain rights."[95]

State Safety Regulations

The well-developed safety regulation of CO_2 pipelines exists on the state level as well. Through delegated authority, states can assume regulatory authority and responsibility for enforcement of HLPSA requirements for intrastate pipelines through certification, whereby states adopt minimum federal standards and make an annual certification to OPS.[96] Alternatively, states can enter into agreements with OPS to oversee aspects of the safety of intrastate pipelines. Also, states can act as agents of OPS with respect to interstate pipelines, such that the state participates in oversight of interstate pipelines but OPS is responsible for the ultimate enforcement in the event of violations.[97]

Both Texas and New Mexico regulate CO_2 pipelines for safety. The Safety Division of the Texas Railroad Commission is certified by OPS to regulate the safety of CO_2 pipelines that are used for intrastate pipeline transportation of CO_2.[98] In addition, the Texas Administrative Code includes safety provisions applicable to hazardous liquids and CO_2 pipelines only.[99] Similarly, New Mexico has a Pipeline Safety Bureau that conducts compliance inspections and investigates accidents involving intrastate CO_2 pipelines. The New Mexico Pipeline Safety Bureau entered into an agreement with DOT whereby the OPS oversees certain aspects of its intrastate hazardous liquids pipelines.

3.3 ADEQUACY OF EXISTING LAWS

Large-scale, commercial implementation of CCS will require not only further development of capture and sequestration technology, but also further delineation of a CO_2 pipeline transportation regulatory regime. This regulatory development will be needed to ensure access to eminent domain to facilitate pipeline construction, and to provide increased regulatory certainty for CO_2 pipeline infrastructure developers that will be necessary for widespread deployment of CCS. While gaps in existing laws and regulations concerning CO_2 pipelines have been discussed above, here they are identified in greater detail:

- *Rate Regulation.* As the CO_2 pipeline network expands, transportation rates could become a significant issue. To date, no federal agency has made an affirmative statement regarding its jurisdiction over CO_2 pipeline transportation rates. Even the STB, which may have jurisdiction to regulate rates under existing law, is limited to interstate pipelines and is sufficiently constrained as to offer little protection to customers. States also have not devoted much attention to rate regulation for intrastate pipelines. Most CO_2 pipelines operate on a contractual basis for a specific application (i.e., EOR).
- *Access Regulation.* Non-discriminatory access could become an important issue as the CO_2 pipeline network expands. While non-discriminatory access is a requirement for receiving a permit under the MLA to cross federal lands,[100]

there is no current regulation of CO_2 pipelines that do not cross federal lands. Policies aimed at avoiding duplication of facilities and capturing economics of scale may impel Congress or the states to impose non-discriminatory access requirements.

- *Safety Regulation.* The current safety regime is well-defined, with PHMSA-administered minimum standards and delegation to states. State programs for CO_2 pipelines are managed by the same agencies that manage other pipeline regulation. This program of delegated authorities on pipeline safety seems to function well in practice, although pressure could grow for some reexamination given recent incidents like the 2010 natural gas pipeline explosion in San Bruno, California.[101] Further build-out of the CO_2 pipeline infrastructure would not appear to require any changes to the existing statutory framework for pipeline safety, so long as the safety regime stays up to date with current pipeline building practices.[102]
- *Siting Authority.* There is currently no federal siting authority for CO_2 pipelines, except on federal lands. Thus, under existing law, pipelines are largely dependent on state eminent domain authority to site both interstate and intrastate CO_2 pipelines, although such authority is not available in all states. As the pipeline network expands (particularly in or through states with no EOR experience), federal siting authority for interstate CO_2 pipelines may become a practical necessity. In addition, the cumbersome process for issuing permits for projects crossing federal lands needs to be streamlined.

3.4 OPTIONS FOR CREATING A CO_2 PIPELINE REGULATORY FRAMEWORK

There are various approaches that could be adopted to regulate CO_2 pipelines. First, the existing model for CO_2 pipeline regulation could be retained. Under that model, states would retain authority for siting all CO_2 pipelines. The federal government only involves itself in siting CO_2 pipelines that cross federal lands. The STB does not regulate rates or access. The authority for pipeline safety would remain within the Department of Transportation, within PHMSA, with state involvement if states so choose.

Second, the model that currently exists for oil pipelines could be used for CO_2 pipelines. Under this model, the states would be responsible for pipeline siting. FERC, rather than the STB, would have authority for transportation rates and access. Safety issues would be handled by PHMSA.

Third, the natural gas pipeline model could be applied. This model envisions a larger federal role. FERC would have authority for the siting of CO_2 pipelines, like the authority provided for natural gas pipelines in the Natural Gas Act.[103] In addition, FERC would be responsible for transportation rates. The authority for pipeline safety would remain within the Department of Transportation, under PHMSA.[104]

In addition, there are other models that could be used for siting of CO_2 pipelines. For example, a federal "backstop" authority, like that provided for electricity transmission siting in EPAct 2005, could serve to keep CO_2 pipeline development on schedule.[105] Under such a model, states would have initial siting authority. However, if a state failed to act and there was a need for such development, FERC would be authorized to issue a permit to developers of CO_2 pipelines.[106] FERC would act to issue permits that would provide federal eminent domain authority to holders of those permits.

Finally, an "opt-in" approach could be used for CO_2 pipeline siting. In this case, the current regime of state siting would continue, but developers of new pipelines could choose whether or not to avail themselves of federal siting authority. Under this approach, CO_2 pipeline developers who need federal siting authority in connection with construction of their CO_2 pipelines could apply for a federal certificate. If granted, the certificate would provide the developer with federal authority to construct and operate the pipeline using federal eminent domain authority, notwithstanding state law. If the Congress were to provide pipeline developers with federal eminent domain authority, it is likely that it would also subject the pipeline to some form of federal economic regulation by FERC or another agency. That regulation could entail non-discriminatory access requirements modeled on the MLA or full rate and service regulations modeled on the NGA.

3.5 RECOMMENDATIONS FOR REGULATING CO_2 PIPELINES

Substantial change in CO_2 pipeline regulation likely will be necessary to support large-scale commercial deployment of CCS that may take decades to complete. It is unlikely that existing state-by-state siting processes and eminent domain authority will be sufficient to support construction—over a period of one or two decades—of a network of interstate CO_2 pipelines perhaps tenfold larger than the current system, or approaching the size of the current natural gas pipeline system. At least some developers will likely need access to a federal siting process and federal eminent domain authority to enable construction of this national CO_2 pipeline system. This authority is likely to be particularly needed for multi-state projects and for projects in states that do not provide CO_2 pipelines with eminent domain authority.

In addition, existing law governing access and rate regulation of CO_2 pipelines is unclear at best. Greater certainty as to the extent of that regulation will help facilitate project financing. In order to obtain financing, project developers (and their debt and equity investors) need to know what regulatory requirements—if any—will apply to the pipeline during its operational phase, so they can evaluate potential regulatory risks.[107] Moreover, if Congress is asked to grant federal siting and eminent domain authority to such pipelines, it is likely to impose some form of "common carrier" requirements, such as non-discriminatory access and rate regulation—among other reasons, to avoid a multiplicity of small high-unit-cost facilities.

Finally, the existing framework for safety regulation of CO_2 pipelines—which relies on a federal regulatory program, with delegation of some functions to state regulators—seems clear and workable.[108]

In light of these considerations, we conclude that Congress should give serious consideration to an "opt-in" federal regulatory regime for new CO_2 pipelines that consists of the following elements:

- The current system of state siting and economic regulation of CO_2 pipelines is retained, except with respect to those new CO_2 pipeline projects for which a permit application is filed, as provided below.
- Any entity proposing to construct a new CO_2 pipeline to transport CO_2 for purposes of permanent sequestration could either proceed under existing state laws or elect to apply to FERC for a federal siting permit for the new pipeline. FERC would have exclusive authority, similar to that under the NGA, to consider and grant or deny the applications. FERC could impose conditions on any permit granted. FERC would undertake environmental reviews comparable to those now conducted under the NGA (see description above).
- Once a FERC permit was granted, the project sponsor would have federal eminent domain authority, and the permit would have the same preemptive effect over state and local land use regulation as a certificate of public convenience and necessity now does under the NGA.
- When operational, the pipeline would be subject to non-discriminatory access and rate regulation similar to FERC's current authority over oil pipelines and the STB's authority over commodity pipelines. Prescriptive regulation of rates and service—on the NGA model—would not, and probably should not, be required.
- Whether or not a pipeline opts for a federal siting permit, it should have access to a streamlined permitting process for use of federal lands.

If Congress is serious about implementing CCS as a GHG reduction strategy, it should address these matters sooner rather than later. Project sponsors will require greater certainty as to the CCS pipeline regulatory ground rules applicable to new CO_2 pipelines in order to support commercial scale CCS projects, which may be ready for deployment as soon as the next decade.

4

PERMITTING GEOLOGIC SEQUESTRATION SITES

A comprehensive legal and regulatory framework for geologic sequestration (GS) is needed to provide structure and certainty for CCS project developers, investors, and the public, and to ensure that injection operations are safe and permanent. This framework should include: 1) permitting rules governing GS site selection and operation; 2) processes for obtaining access to pore space; 3) arrangements for management of long-term liability and stewardship; and 4) greenhouse gas accounting protocols to link CCS to future climate or energy programs. This chapter focuses on regulation of GS site permitting. Policy governing access to pore space, long-term liability issues, and greenhouse gas accounting is covered in Chapters 6, 7, and 8, respectively.

Overall, the objectives of site permitting rules should be to insure that GS sites are carefully selected and responsibly operated. To be successful, GS permitting rules should:

- demonstrate that a proposed site can accept and contain carbon dioxide (CO_2) at a commercial scale;
- comprehensively manage potential risks to health, safety, the environment, and property;
- facilitate public engagement;
- provide coordination to insure that the quality of GS site operations is consistent nationwide and that interstate projects can operate efficiently;
- be reasonably consistent with regulation of closely related activities, such as the disposal of waste fluids and CO_2 injection for enhanced oil recovery (EOR);
- include mechanisms to balance multiple environmental objectives; and
- specify conditions for site closure and transition to long-term stewardship.

The EPA finalized permitting rules for GS sites in December 2010 under the Underground Injection Control (UIC) program.[1] These rules accomplish portions of the objectives outlined above, but due to restrictions in the UIC program's statutory mandate and historical artifacts of how the program has evolved, the new rules fall short of creating comprehensive GS site permitting regulation. A number of states have also passed laws and adopted rules regarding GS site permitting. While some of these state-level permitting rules are more comprehensive than the federal UIC rules, none meet all the objectives outlined above, and furthermore they differ substantively from state to state.

Section 4.1 describes the UIC program and details the new federal GS site permitting rules (UIC Class VI). Section 4.2 discusses state-level regulations on GS site permitting. Section 4.3 discusses the importance of public engagement. Section 4.4 presents recommendations to make the regulatory framework for GS site permitting more comprehensive and conducive to large-scale commercial CCS.

4.1 THE US EPA UNDERGROUND INJECTION CONTROL PROGRAM

The UIC program is one of a number of programs created after the passage of the Safe Drinking Water Act (SDWA) of 1974.[2] The purpose of the SDWA is to protect the quality of drinking water in the United States including above-ground and underground waters currently or potentially used for drinking water. The role of the UIC program, which began in 1979 and first promulgated regulations in the 1980s, is to protect Underground Sources of Drinking Water (USDWs) regardless of depth of the water bearing formation and independent of whether the water is actually in use as a source of public drinking water.[3] Under the UIC program, the EPA (or states to which the EPA has delegated authority giving them "UIC primacy") has the authority to exempt some USDWs, thus allowing injection where it would otherwise be prohibited. The Agency explains that an exempted aquifer is:

> An aquifer, or a portion of an aquifer, that meets the criteria for a USDW, for which protection under the SDWA has been waived by the UIC Program. Under 40 C.F.R. Part 146.4, an aquifer may be exempted if it is not currently being used—and will not be used in the future—as a drinking water source, or it is not reasonably expected to supply a public water system due to a high total dissolved solids content. Without an aquifer exemption, certain types of energy production, mining, or waste disposal into USDWs would be prohibited. EPA makes the final determination on granting all exemptions.[4]

UIC Program Structure

From 1980 through 2009, the UIC program included five classes of wells. A sixth well class was added in 2010 for wells injecting carbon dioxide for geologic sequestration.

- Class I wells inject hazardous waste, industrial non-hazardous waste, and municipal wastewater. Nationwide, there are approximately 550 Class I wells.[5] Construction, monitoring, and reporting requirements are generally the most stringent for Class I hazardous wells. All Class I wells must inject below the deepest drinking water aquifer.
- Class II wells are associated with oil or gas production. They include disposal wells, enhanced recovery wells, and hydrocarbon storage wells. There are approximately 144,000 Class II wells in the US.[6] Wells that are injecting CO_2 for EOR fall in this class.
- Class III wells are used for solution mining of minerals below the deepest USDW aquifer. There are approximately 18,500 Class III wells.[7]
- Class IV wells inject hazardous or radioactive wastes above or into a drinking water aquifer. They are banned unless specially authorized as part of a groundwater remediation project. Nationwide, thirty-two sites use Class IV wells.[8]
- Class V includes all other kinds of injection wells. Class V wells typically inject non-hazardous waste into shallow formations. There are 400,000 to 650,000 Class V wells—the exact number is not known.[9] In March 2007, the EPA issued guidance for pilot project geologic sequestration wells to be permitted as Class V, experimental wells.[10]
- Class VI includes wells injecting carbon dioxide for the purpose of geologic sequestration. As of the middle of 2011, no wells have been permitted as Class VI although an application had been submitted by Archer Daniels Midland[11] and others were in preparation.

For each injection well class, the EPA UIC rules establish minimum requirements for permitting, operations, and reporting. Regulatory oversight can be either by regional EPA offices, or by state regulators. In much of the country, states are central to implementation and enforcement of UIC rules. States apply for primary enforcement authority by demonstrating that the state UIC program is:

> *at least as stringent* as the Federal standards; the state, territory, or tribal UIC requirements may be more stringent than the Federal requirements. (For Class II, states must demonstrate that their programs *are effective* in preventing pollution of USDWs.)[12]

The EPA may grant a state primacy for the entire UIC program, or for just certain classes of injection wells. Currently, thirty-three states have full primacy, seven share

FIGURE 4.1 *The six well classes that currently exist under the federal/state Underground Injection Control (UIC) program. Figure adapted from USEPA.*

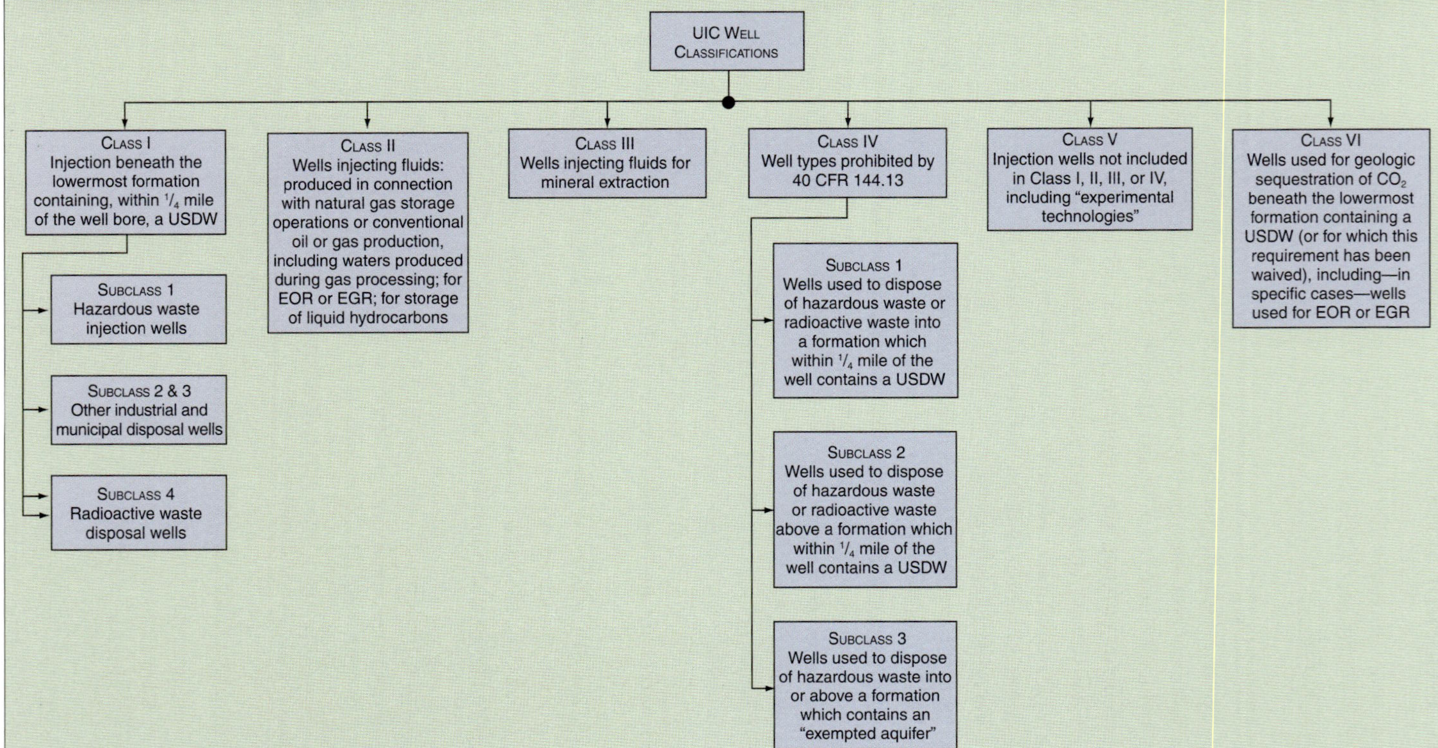

authority with the EPA, and ten have UIC programs run by the EPA, as shown in Figure 4.2. States with existing regulatory primacy will need to apply for primacy to regulate the new Class VI wells.

GS Site Permitting Under UIC Class VI Rules

The US EPA promulgated a new rule for GS wells in December 2010 that creates the Class VI well category.[13] Class VI rules apply only to wells that inject CO_2 for geologic sequestration. Wells injecting CO_2 for EOR remain Class II wells. The Class VI rules require that all GS wells inject beneath the deepest USDW. They outline construction requirements for GS wells similar to those for Class I hazardous waste disposal wells, including use of materials compatible with the CO_2 stream. Class VI rules require that a permit applicant submit:

- information on the site geology and hydrogeology;
- proposed operating conditions;
- a review of all wells in the area of review (AoR) that penetrate the injection or confining zone, and plans for corrective action on any wells that could serve as conduits for CO_2 leakage;
- injection well testing plans;
- monitoring plans;

Figure 4.2 *In much of the US, states are responsible for administration of the UIC program (green). In the case of six states (red), implementation is jointly handled between the state and the EPA. In the balance of the country (blue), the EPA administers the program directly. (Figure from US EPA.)*

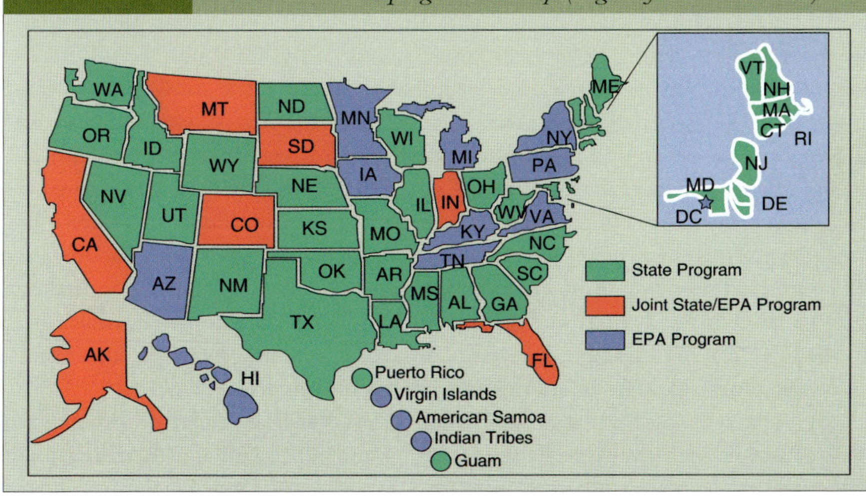

- emergency and remedial response plans;
- plans for plugging the injection wells when injection ends;
- post-injection site care and site closure plans;
- proof of financial responsibility and resources for corrective action, injection well plugging, post-injection site care, and site closure.

As noted below, this list does not include a number of issues essential to the long-term viability of commercial-scale CCS.

The rules specify that EOR operators holding UIC Class II permits who want to transition to injecting CO_2 primarily for geologic sequestration must apply for a Class VI permit if there is an increased risk to USDWs.[14] This determination will be based on factors such as changes in reservoir pressure, increase in CO_2 injection rates, decrease in hydrocarbon production, and the source and properties of the CO_2 to be injected. The rules also effectively limit the use of aquifer exemptions for Class VI wells to circumstances where sites transition from EOR to GS.[15]

Requirements for Class VI wells differ in several important ways from requirements for other well classes. Generally they are more stringent, but they also begin to incorporate some more contemporary regulatory practices, such as the use of adaptive and performance-based regulation, as discussed further in Chapter 5. They are most similar to requirements for Class I hazardous waste wells, and significantly different from requirements for Class II wells that inject CO_2 for EOR. Many of the differences described below can be traced to the historical development of the UIC program, and to the fact that expectations of environmental regulations have evolved considerably in the thirty years since the rules for UIC Well Classes I through V were written.

One difference is in the size of the AoR, which is defined as the area that may be impacted by injection. For other well classes, the AoR is a fixed radius around a well (for example, 0.25 miles for UIC Class I non-hazardous wells), but for Class VI the EPA requires that the extent of this region be calculated by numerical modeling of the maximum extent of the CO_2 plume and the area where pressure differentials caused by injection could cause movement of fluids into a USDW.[16] This definition means that the AoR for Class VI wells will be much bigger than for other well types. The size of the AoR determines the areal extent of site characterization studies and the size of the area where the operator must plug other wells that could create a pathway for injected fluids to migrate into USDWs (termed corrective action). For other well classes, corrective action must be completed on the entire AoR before injection begins, but Class VI permits will allow phased corrective action. This approach allows an operator to fix wells closest to the injection site first, and then gradually work outward, always staying ahead of the expanding CO_2 plume and induced brine movements.

Site characterization requirements for Class VI wells differ from injection well classes in both the size of the area that must be studied and their stringency. For example, a permit application for a Class II CO_2 EOR well in Texas is a one-page form that requires only the proposed depth of the well, data about the injection formation such as porosity and permeability, and the locations of existing wells within a half-mile.[17] A permit for a Class VI CO_2 injection well requires a detailed site characterization study, estimated to cost approximately $2 million.[18]

Monitoring requirements are also more extensive for Class VI wells than for other injection well classes. GS sites are required to use both direct and indirect monitoring methods to measure groundwater quality, CO_2 plume location, and subsurface pressure perturbations.[19] Federal UIC rules for other UIC well classes require little to no direct or indirect monitoring.

Post-injection requirements are another area where rules for Class VI wells are more stringent than for other injection well types. Class VI rules require that after injection is finished, the operator must continue to monitor the site and provide financial assurance until the injected CO_2 is stable and poses no risk. The post-injection period is fifty years, but the EPA has discretion to adjust the length based on site performance. Permits for Class I hazardous waste wells can require limited post-closure monitoring; however, for most other well classes, the only requirement after injection activities are completed is to submit a plugging and abandonment report.

Compared to past UIC regulations, the Class VI rules are significantly more adaptive and incorporate some of the advances in modeling and monitoring that have emerged in the thirty years since the first UIC rules were promulgated. For example, they allow for a staged development of a GS site, with GS project operators able to perform corrective action in phases, and they require that operators periodically reevaluate the AoR, comparing modeling projections with monitoring results. However, as the other well classes have not been similarly updated to reflect the advancements in modeling, monitoring, and construction technologies, these features make the Class VI rules disproportionately stringent. The contrast between the two well classes that can inject CO_2, Class II and Class VI, is particularly stark. Given the additional cost and stringency of the new Class VI wells, it is possible that the new regulations will preferentially steer GS project developers to develop Class II EOR wells in lieu of GS wells in saline formations.

Regulatory Gaps in Federal GS Site Permitting

The UIC Class VI rules are designed to prevent geologically sequestered CO_2 from harming USDWs. However, in the preamble to the Class VI rule, the EPA notes that because the statutory authority for the UIC program comes from the SDWA, the agency is unable to address other critical issues that must be resolved for geologic sequestration to be successful on a large scale.[20] These include:

- potential human health or ecological impacts if carbon dioxide leaks to the surface or near-surface;

- balancing risks to groundwater with future environmental and human health threats from climate change;
- access to pore space (discussed in Chapter 6);
- long-term liability (discussed in Chapter 7); and
- greenhouse gas accounting (discussed in Chapter 8).

The EPA argues that by protecting drinking water it will effectively regulate other risks because in nearly all geologic situations, groundwater stands between the sequestration reservoir and the surface.[21] But the fact that risk assessments carried out for UIC Class VI site characterization studies will not characterize the risk of surface leakage may prove to be a barrier to local public acceptance of proposed GS sites.

Balancing multiple environmental objectives during the GS site permitting process is an important, yet difficult, consideration. Because the SDWA places protection of drinking water aquifers (independent of quantity or depth) above all other objectives, it cannot address the tradeoffs between risks to groundwater and risks from climate change. To a certain extent, the SDWA authorizes use of aquifer exemptions to balance tradeoffs between energy supply and drinking water protection, but the Class VI rules preclude the use of aquifer exemptions for GS sites in saline formations,[22] which will provide the vast majority of storage capacity for any future commercial-scale CCS. The tension between local versus global interests and present versus future interests makes balancing multiple environmental objectives challenging for regulators and local communities, but experiences such as the UIC program's handling of wastewater injection in Florida illustrate the importance of having balancing mechanisms in place (Box 4.1).

Box 4.1 *Florida Experience with Municipal Wastewater Injection Illustrates need to Balance Multiple Environmental Objectives*

Florida's experiences with the injection of municipal wastewater under the UIC program offer a number of notable lessons. Despite the procedural requirements for Class I well construction and operation, injected municipal wastewater has migrated upward into overlying drinking water aquifers. Because there were no other economically or environmentally viable options for disposing of the wastewater, the EPA promulgated new regulations that allowed injection to continue if advanced wastewater treatment was implemented.[23] In effect, they balanced the objective of environmentally sound wastewater disposal with the objective of groundwater protection. This situation, where the Agency was forced to develop an exception after the fact, should be avoided for GS, by creating balancing mechanisms within the permitting rules. The Florida experience also highlights the value of monitoring: Florida detected the problem because it has monitoring standards for UIC Class I wells that are more stringent than most other states and more stringent than federal UIC rules.

Another regulatory gap, not related to the lack of statutory authority, is the absence of mechanisms to resolve competing applications for a specific GS site location. This could become an issue as CCS deployment reaches commercial scale. We discuss this issue at greater length in Chapter 6.

4.2 THE ROLE OF THE STATES

In much of the country, states will play central roles in regulation of GS. In addition to the fact that states may apply for primary UIC enforcement authority, as discussed above, the regulatory context for geologic sequestration is shaped by state environmental and natural resources law as well as a strong history of state and local control over land-use and property rights issues. State regulators have valuable knowledge specific to their locale. They should and will play an important role in permitting GS projects. Laws and rules related to GS have been passed in a number of states, as summarized in Table 4.1, and described in Box 4.2.

Unlike the federal UIC Class VI rules, some state-level GS laws and regulations also cover property rights and liability issues. Major property rights issues addressed include: pore space ownership, ownership of injected CO_2, whether CO_2 is considered a commodity (as opposed to a waste), liability during injection operations, whether mineral rights have primacy over pore space rights, authority to use eminent domain for GS projects, utility authority and processes, and whether EOR will be subject to GS rules. Some states also address long-term liability and stewardship of GS sites. Laws and rules related to access to pore space (Box 4.3) and liability (Box 4.4) have been passed in a number of states.

Comparison of these early state GS regulations shows that while they agree in some areas, they disagree on some important issues. Most take the position that: 1) protection of drinking water should be paramount; 2) pressure effects from CO_2 injection are important; 3) monitoring should include testing in the injection zone and the first permeable formation overlying the confining zone; and 4) site performance should be a factor in determining the length of the post-injection period. Emerging regulatory discrepancies are the allowable composition of the CO_2 stream, the size of the AoR, reservoir performance goals, and management of risks other than those to groundwater. As yet, it remains unclear how state-level GS regulatory initiatives will fit into a national GS regulatory framework. The longer they proceed without a comprehensive national approach, the more complex and heterogeneous the rules are likely to become.

TABLE 4.1 *Summary of state geologic sequestration regulatory initiatives as of March 2011.*

State	Property Rights, Including Access to Pore Space	Permitting Rules	Long-term Stewardship
Kansas 2010: HB 2418 2009: KAR 82-3-1100-1120 2007: HB 2419	N/A	Agency: KS Corporation Commission. Rules adopted Feb. 2010.	State will assume long-term site management. Fund established for long-term monitoring and remediation. State assumes no liability.
Louisiana 2009: HB 661 2008: HB 1220, 1117	Addresses CO_2 ownership, liability during operations, and eminent domain. Does not address pore space ownership.	Agency: Office of Conservation, Dept. of Nat. Resources. Rules not yet proposed.	State will assume long-term ownership but limited liability. Fund established for long-term monitoring and limited remediation.
Montana 2009: SB 498	Addresses pore space ownership, liability during operations, mineral rights primacy, CO_2 ownership, and provides for unitization.	Agency: MT Board of Oil and Gas Conservation, with comments from MT Board of Env. Review. Rules not yet proposed.	State will assume long-term ownership and liability. Fund established for all long-term liabilities.
North Dakota 2009: SB 2095, SB 2139	Addresses pore space ownership, CO_2 ownership, liability during operations, mineral rights primacy, and provides for unitization.	Agency: ND Industrial Commission. Rules adopted November 2009.	State will assume long-term ownership and liability. Fund established for all long-term liabilities.
Oklahoma 2009: SB 610 2008: SB 1765	Addresses CO_2 status and ownership and mineral rights primacy. Inventory accounting rules adopted. Does not address pore space ownership and liability during operations.	Agency: Corporation Commission for fossil-fuel-bearing formations, Dept. of Env. Qual. for all others. Rules not yet proposed.	N/A

Texas 2009: HB 1387, HB 1796	Addresses CO_2 ownership, liability during operations. Does not address pore space ownership.	Agency: Railroad Commission, Texas Commission on Env. Quality. Rules adopted November 2010.	State will assume limited long-term site management but not all liabilities. Fund established for long-term monitoring and limited remediation.
Utah 2008: SB 202	N/A	Agency: Dept. Of Env. Quality. Rules not yet proposed.	N/A
Washington 2008: WAC 173-407-110 2007: ESSB 6001	N/A	Agency: Department of Ecology. Rules adopted in 2008.	N/A
West Virginia 2009: HB 2860, W.V. Code, Chap. 22, Art. 11A	Addresses mineral rights primacy. Assigns study group to make recommendations on other issues such as pore space ownership by 2011.	Agency: Dept. of Env. Protection. Rules not yet proposed.	N/A
Wyoming 2010: HB 17 2009: HB 57, 58, 80, Water Qual. Rules & Regs. Chap. 24 2008: HB 89, 90	Addresses pore space ownership, CO_2 ownership, liability during operations, mineral rights primacy, and provides for unitization.	Agency: Dept. of Env. Quality. Rules adopted November 2010.	State will assume limited long-term site management but not all liabilities. Fund established for long-term monitoring.

Box 4.2 *State Policies on GS Site Permitting*

As of March 2011, five states (Kansas, North Dakota, Texas, Washington, and Wyoming) have issued GS permitting rules, and five more (Louisiana, Montana, Oklahoma, Utah, and West Virginia) have passed laws initiating the process. These state rules could serve as the basis for state implementation of the UIC Class VI program, but significant differences have emerged in their approaches. For example, in Kansas and North Dakota, GS is treated as a unique activity under the authority of stand-alone GS legislation and the state will permit sequestration facilities as opposed to permitting just CO_2 injection wells.

As with the federal level, state-level policy choices are made in two places: 1) in authorizing legislation; and 2) during rulemaking proceedings implementing the legislative directive. At the legislative level, an important policy choice is which type of agency will be directed to draft the GS regulations (i.e., an agency charged with environmental protection or an agency charged with managing natural resource development (such as oil and gas regulators)). In Kansas, Louisiana, and North Dakota, the chosen agencies are natural-resources-based. Utah, Washington, West Virginia, and Wyoming chose their environmental protection agencies.

Montana, Oklahoma, and Texas chose hybrid approaches. Montana directed the Board of Oil and Gas Conservation to develop GS permitting rules with substantial input from the Board of Environmental Review. Oklahoma divided jurisdiction based on the nature of the potential GS site, putting the Corporation Commission in charge of GS in fossil-fuel-bearing geologic formations and its Department of Environmental Quality in charge of GS in saline formations. Texas divides jurisdiction between its Commission on Environmental Quality for offshore GS and its Railroad Commission for onshore GS sites.

Once the policy choice is made to determine which state agency will develop regulations, additional policy choices are made at the agency level. At the state level, these rules continue to evolve. For example, Washington State implements rules broader than the US EPA's Class VI well rules. They are coupled with a strong state climate policy, which permits the state's environmental protection agency to link GS with greenhouse gas reductions, whereas the federal rules are limited to the protection of drinking water.[24]

Wyoming's proposed rules are primarily adopted from existing UIC regulations for Class I, Class V, and Class VI wells. Wyoming does, however, include some additional language that requires well owners, as part of a monitoring plan, to "assess the migration of the injected carbon dioxide, and ensure the retention of the carbon dioxide in the geologic sequestration site." Such language is absent from the federal rules.

One of the most striking aspects of state GS permitting laws and rules is that they uniformly exempt EOR. In fact, the case has been made that one of the primary outcomes of state policy on GS has been to protect EOR business as usual.[25] Several states (MT, ND, KS) have also addressed how to convert EOR injection sites to GS injection sites, or have directed a state agency to develop rules to govern that process.

Box 4.3 *State Policy on Access to Pore Space*

Subsurface property rights are traditionally governed by the states, but there are no uniform rules for subsurface use rights among and between the states. Three states—Montana, North Dakota, and Wyoming—have passed laws declaring that surface owners own pore space rights.[26] North Dakota's statute prohibits severance of pore space rights from surface rights. These three states have also created unitization procedures for GS, providing developers with a mechanism to obtain access once 60% (MT and ND) or 80% (WY) of owners consent to the unitization plan. Table 4.1 gives additional information about state policy on property rights issues.

Box 4.4 *State Policy on GS Liability Issues*

A number of states have passed laws that address GS liability, during both operations and long-term stewardship. Liability during operations is generally assigned to the site operator, through laws specifying that he or she owns the injected CO_2. The operator is free to assign those rights with a legally binding instrument (e.g., a contract or a deed). Montana, North Dakota, Oklahoma, Texas, and Wyoming all take this approach. Louisiana leaves the issue of CO_2 ownership to the interested parties to determine by contract without creating any default presumption for the operator. Texas created two special cases where a state entity would own the CO_2 (FutureGen and sites off shore on public land). For other GS projects, the operator owns the CO_2 by default in Texas. Illinois also passed a law that pertains directly to the FutureGen project, designating the state to assume any liabilities associated with the project's sequestered CO_2.

Six states have taken action regarding long-term liability (KS, LA, MT, ND, TX, WY). They have established funds dedicated to long-term stewardship of GS sites, and at least partially addressed assignment of long-term liability. States' intentions regarding long-term liability appear in legislation as statements regarding: 1) states' long-term responsibilities; 2) operators' long-term responsibilities; and 3) authorized uses for money from the GS fund.

These six states cover a wide spectrum in the amount of long-term liability they have agreed to accept. The following table details both what states have agreed to take on, and what they have specified that operators shall be released from, with the states listed in order of increasing assumption of long-term liability. Wyoming would only take over long-term monitoring,[27] whereas Texas would also perform certain types of remediation,[28] and Kansas would perform any needed remediation.[29] Louisiana's provisions raise questions in that the state would take ownership of closed GS sites and assume responsibility for long-term monitoring and certain types of remediation, but the law simultaneously releases the operator from all long-term liability while saying that the "state shall not assume or have any liability by the mere act of assuming ownership of a storage facility."[30]

Similarly, Kansas and Wyoming declare that the state assumes no liability. North Dakota and Montana go the farthest. Both would assume all rights and interests in and all responsibilities associated with the closed GS site.[31] Montana goes further than North Dakota in that it would release operators not just from regulatory responsibility, but also from all liability associated with the closed GS site. North Dakota and Montana also both add the caveat that the state will be responsible for long-term GS site monitoring and management until such time as the federal government takes responsibility. Overall, state laws on long-term liability leave major gaps and significant questions. Table 4.1 gives additional details on state policy on liability and property rights.

Long-term Liability for Geologic Sequestration Sites

State assumption of long-term liability: Less (top) ⟶ More (bottom)

State	State takes over. . .	Operator released from. . .	State "assumes no liability"
Wyoming	Long-term measurement, monitoring, and verification of geologic sequestration sites.	Nothing.	Yes
Texas	Long-term monitoring, remediation of mechanical problems, repairing mechanical leaks at geologic storage facilities.	Nothing.	
Kansas	Any future remediation or monitoring activities.	Nothing.	Yes
Louisiana	Long-term monitoring of the site, including remediation of any well leakage.	All (regulatory) duties and all liability associated with or related to that storage facility which arises after the issuance of the certificate of completion.	Yes
North Dakota	All rights and interests in and all responsibilities associated with the stored carbon dioxide.	All regulatory requirements associated with the storage facility.	
Montana	All rights and interests in and all responsibilities associated with the geologic storage reservoir and the stored carbon dioxide.	All regulatory requirements and liability associated with the geologic storage reservoir and the stored carbon dioxide.	

4.3 COMMUNITY ENGAGEMENT DURING THE PERMITTING PROCESS[32]

Successful commercial deployment of CCS will be at least partially contingent on local community acceptance of individual GS projects. While this is true for all large infrastructure developments, it is especially critical in the case of GS because there is a lack of awareness and understanding of GS among the general public,[33] and host communities of existing pilot-scale and commercial-scale GS projects have been found to raise more concerns and questions than those of more established industries and technologies.[34] In the absence of a concerted outreach effort, recent experience suggests that community members will form their own (sometimes mis- or un-informed) opinions of GS.[35] Since pervasive misconceptions can be hard to change,[36] it is important to proactively engage and educate host communities. When people are more informed about a technology, they tend to show more support for it,[37] and recent work suggests that public acceptance of CCS is high when careful detailed information is provided and it is viewed as a part of a portfolio of low carbon energy strategies.[38] Providing the community with early access to technically accurate and understandable information about the GS project will likely also increase trust in project developers and regulators. Gaining and maintaining trust can play an integral role in public risk perceptions.[39] Additionally, respectful treatment of the local community will be critical to national public acceptance and the long-term success of GS technologies.

The EPA recognized the importance of involving the public in permitting decisions in the preamble to the UIC Class VI rule, and acknowledged that early and frequent public education and participation would be particularly beneficial for GS projects.[40] However, the Class VI rules essentially follow existing UIC requirements on public participation,[41] rather than making modifications to reflect the unique nature of GS, the advent of new information technology, or current understanding of best practices in community engagement. The only new provision, unique for Class VI wells, is a requirement to notify additional state agencies such as state and local oil and gas agencies, agencies that regulate mineral resources, the Public Water Supply Director, and other agencies that oversee injection wells.[42] Given the public's unfamiliarity with GS technology and issues, this standard approach will likely fall short of a comprehensive public engagement effort. GS permitting rules should require modern methods of risk communication,[43] just as they require modern methods of site characterization and monitoring.

Best practices for community engagement are described in a recent report by an expert working group convened by the World Resources Institute.[44] They recommend that a deliberate process of communication among project developers, community members, and regulators should start during the permitting process and continue throughout the project life cycle. We believe that key elements of these guidelines should be incorporated into GS permitting rules, to insure that all GS projects plan for community engagement, just as they plan for well testing, monitoring, and remediation. This could be accomplished using a two-tiered approach that makes expectations clear for both project developers and regulators. First, development and approval of a site-specific community engagement plan should be a requirement in order to obtain a Class VI permit. Second, the permitting agency should communicate with the local community about GS and about the permitting process.

The community engagement plan should outline systematic, site-specific procedures that the project developer will follow throughout the GS project life cycle, including but not limited to:

- an assessment of the local community—their needs, concerns and any special social consideration that should be included in the engagement process;[45]
- an explanation of the reasons for the proposed project and the alternatives that have been considered;
- a plan to ensure that the community has easy access to information, including a strategy for the distribution of information about the project to the local community, other stakeholders, as well as the media to be used and the timing of information releases (i.e., during the permitting process, CO_2 injection and post-closure phases);
- a procedure for addressing community concerns (through venues such as public meetings, mediated discussions or other less formal measures);
- an overview of the information content that will be provided to the local community, addressing the costs, benefits, and risks of the project, as well as mitigation, emergency management, and contingency plans;
- a plan to reevaluate and revise the community engagement plan as necessary.

The permitting agency should also play a role in communicating to the local community about the permit approval process and GS site operations and risk management, using a communication strategy that includes, but is not limited to:

- basic information about GS, existing GS projects, and its role in climate change mitigation (including a comparison to other low-carbon technologies);
- information about the permitting and risk mitigation processes, as well as about the enforcement of regulations over GS project developers;
- information describing how members of the general public or the local host communities can voice concerns (through public hearings, mediated discussions, etc.).

4.4 RECOMMENDATIONS FOR PERMITTING GS SITES

The current regulatory framework for GS site permitting created by the federal UIC Class VI rules (supplemented by some state rules) provides a strong start in managing GS risks, but it falls short in several important areas, largely because the UIC program is only authorized to address risks to drinking water. Thus, current permitting rules are unable to comprehensively manage environmental, health, and safety risks, or to balance multiple environmental objectives. They are also unable to address access to pore space or to assure adequate management of long-term stewardship. Creation of a comprehensive legal and regulatory framework that can facilitate large-scale deployment of CCS will require new federal legislation, amendments to existing legislation, and modifications to existing rules. Here we present recommendations for GS permitting rules that can more fully support commercial CCS deployment. The model legislation in the Appendix illustrates how we believe these recommendations should be implemented.

Chapters 6 and 7 present recommendations related to pore space access and long-term liability. In Chapter 5, we present a plan for regular review and revision of GS rules, which could serve as a vehicle for introducing these recommended changes.

Federal legislation should declare that sequestering carbon dioxide (CO_2) in geologic formations to mitigate the detrimental effects of climate change is in the public interest. This will lay the foundation for SDWA amendments to comprehensively manage risk and balance multiple environmental interests.

Legislation should amend the SDWA to authorize UIC regulators to assess and manage all potential risks to human health or the environment from GS. Class VI permitting rules should require that all potential risks to human health or the environment be taken into account in site characterization studies, community engagement plans, operating conditions, monitoring plans, and remedial response plans. Having comprehensive GS risk management oversight housed with one regulator is more efficient for both public and private actors, and will facilitate more transparent community engagement.

Legislation should amend the SDWA to authorize regulators to balance protection of underground sources of drinking water (USDW) with other compelling national interests, such as mitigating climate change. As with so many other infrastructure and energy projects, the challenges in balancing the greater social welfare with the vested, and often vocal, interests of those directly impacted will be difficult, but the UIC program has experience in making these tradeoffs (for example, with aquifer exemptions that allow hydrocarbon production, and with the handling of municipal wastewater injection in Florida). While this proposed amendment would likely prove controversial, it simply makes overt the often unacknowledged fact that regulators are often called on to balance multiple environmental objectives.

UIC Class VI rules should be revised to require a community engagement plan. The best approach for this dialogue will be as site specific as all other aspects of a GS project, because the needs, experiences, and context of each community is unique. But constructive relationships with local communities will be crucial for commercial deployment of GS, so development and approval of a community engagement plan should be made a condition for obtaining a Class VI permit.

UIC rules should be revised so they can weigh competing applications for Class VI permits. Details on how this might best be achieved are discussed in Chapter 6.

5

LEARNING FROM AND ADAPTING TO CHANGES IN GEOLOGIC SEQUESTRATION TECHNOLOGY

New and emerging technologies pose challenges for regulation. Regulators cannot fully anticipate the entire scope of a technology's impact or predict the ways in which the technology will change over time. Even the most comprehensive regulations may inadvertently fail to cover important contingencies or be overly confining and hamper future developments. As a technology advances and the knowledge base expands, the associated regulatory system should be able to evolve in response to new developments. The fact that implementing an adaptive approach often has proved to be difficult in the past means that considerable care should be taken when designing new regulatory frameworks.

Geologic sequestration (GS) is an ideal candidate for a regulatory system that is capable of evolving and adapting, because it is a young technology with many unknown qualities. Decades of experience with analogous activities, such as enhanced oil recovery, natural gas storage, and underground waste injection, provide a solid foundation from which to begin to craft regulations for geologic sequestration. However, scientific, technical, and regulatory communities have limited experience with the specific challenges of injecting millions of tonnes of carbon dioxide into deep geologic formations. Much will be learned as the first commercial-scale GS projects are built and operated. Therefore, GS regulations will be most effective if regulators can learn from and adapt to changes in GS technology and modify the regulations in response.

Two concepts hold promise for keeping regulation effective and efficient as more is learned about geologic sequestration:

- *Performance-based regulation* focuses on results, rather than on the means by which they are achieved. It is an alternate approach to design- or technology-based regulation that specifies particular behaviors, technologies, procedures, or processes.[1] The goal of performance-based regulation is to offer certainty to both the permit holder and the public regarding performance expectations for compliance, along with flexibility for the permit holder to meet those expectations in the manner they choose. As a result, compliance strategies can evolve as knowledge evolves, while the required level of performance remains the same.

- *Adaptive regulation* is a deliberate process of periodic review and revision that uses experience and new knowledge to clarify and improve the rules. Adaptive regulation is not retroactive; it does not change the regulatory requirements for existing projects unless all affected parties agree that it should. It can, however, adapt the rules so new permits are granted based on the best current scientific and technical knowledge.

The following sections describe performance-based regulation and adaptive regulation in more detail, including examples of programs where these approaches have been used and critiques of current GS regulations. The chapter concludes with our recommendations for a regulatory framework for geologic sequestration that is performance based and includes a prescribed process for periodic review and revision.

5.1 PERFORMANCE-BASED REGULATION

Performance-based regulation is defined as regulation that specifies the required outcome, not the means to achieve it. It is an alternative to regulation based on design or technology standards. In practice, the distinction is not always clear. There are examples of real-life regulations that fall along a continuum, from purely performance-based regulation at one end to purely prescriptive regulation at the other.[2] Traditionally, the majority of environmental regulations have fallen toward the prescriptive, or "command and control," end of this spectrum. However, recent decades have witnessed a push to make regulation more performance based as a means of reaching specific regulatory objectives while encouraging technical innovation and decreasing the cost of compliance.[3] For example, the Clinton administration's strategy for reinventing environmental regulation stated that "Environmental regulations must be performance-based, providing maximum flexibility in the means of achieving our environmental goals, but requiring accountability for results."[4]

Creating effective and workable performance-based regulation does present some difficult challenges. For example, as performance options grow in complexity, regulators must be well versed in the technical details in order to oversee them effectively. In addition, a balance must be struck between regulatory discretion and control, such that regulators have enough discretion to promote flexibility and creativity, but not so much that they become vulnerable to capture by the regulated industry.[5]

Performance-based regulation has three essential components:[6]

1. Characterization of outcomes—for example, the creation of regional air quality goals that prevent unhealthy conditions.
2. Standards for the desired levels of performance—for example, air quality thresholds based on scientifically documented health effects.
3. Assessment procedures for gauging the level of performance—for example, physical measurements of emission levels.

The ideal of flexible, performance-based regulation is hard to fault, and the first of the necessary components, the desired outcome, usually is not hard to articulate. In the case of geologic sequestration, the desired outcome is cost-effective reduction of greenhouse gas (GHG) emissions without damage to health, safety, the environment, or property. The second two components needed for successful performance-based regulation, however, present greater challenges. Establishing standards to evaluate GS performance and agreeing on methods to demonstrate compliance are technically demanding tasks, subject to both ambiguity and uncertainty.

Performance-based regulation can create room for innovation and learning, but it can present challenges for regulators who are typically more familiar with using standardized rules and applying them in a routine manner. It typically requires regulators to have more substantive technical expertise and to exercise more creative judgment than has been common in the past. In addition, since individual firms may pursue a variety of compliance solutions, regulators may need to understand and assess a number of different approaches. Tension between regulator and regulated exists under all forms of regulation, in large part because regulated entities cannot be supervised at all times, and because regulators do their job with limited information and resources. Performance-based regulation has the potential to exacerbate this asymmetry of information, unless regulators can stay current with rapidly advancing technical information. In the end, the potential benefits of performance-based regulation—flexibility, innovation, the ability to incorporate new knowledge, and decreased compliance cost—can only be achieved if the public is willing to make the investment in regulatory capacity.

5.2 EXAMPLES OF PERFORMANCE-BASED REGULATION

Performance-based regulation is not a new concept; it has been used in the United States and other countries in a number of areas, including air and water quality, energy efficiency, consumer product safety, food safety, pipeline safety, and building safety.[7] The following examples of price-cap regulation of electric utilities and building codes in New Zealand illustrate the importance not only of characterizing desired outcomes, but also of setting appropriate performance standards and identifying workable assessment procedures.

Regulation of Electric Utilities

Performance-based regulation for utility rates gained popularity in the 1990s. The desired outcome is that consumers receive good service for a set price. Performance standards include a price cap that cannot be exceeded by the utility company (except to keep up with inflation), and quality measures such as an allowable number of service interruptions.[8] The firm accepts the risks associated with locking into a set rate for a predetermined period of time, but can profit by being innovative; if it finds ways to provide the required service at a cost below the price cap, the financial savings are returned to the firm, or in some schemes, split between customers and the firm. Depending on the arrangement, failure to achieve the performance standard can result in penalties.[9] Assessment procedures are quite straightforward, because public utility commissions are well established and organized for this type of monitoring. With clearly characterized outcomes, well-defined performance standards, and straightforward assessment procedures, price cap regulation of electrical utilities has been quite successful in achieving its objectives.

Building Codes in New Zealand

New Zealand's experience with performance-based regulation of building codes points to the problems that can arise if performance-based rules are not backed by appropriate performance standards and capable assessment.

In an effort to make regulation less burdensome and to encourage innovation and the adoption of cost-effective practices, regulatory reformers in New Zealand passed the Building Act of 1991, which instituted performance-based building codes.[10] Within a decade, these well-meaning reforms resulted in the "Leaky Building Crisis," which became a major issue for the central government as fundamental failings in tens of thousands of newly constructed homes became apparent. By 2010, costs for repairs were estimated at from US$7 to 15 billion.[11]

Subsequent investigations of what went wrong found no fault with the general concept of performance-based regulation, but concluded that New Zealand's building code had failed to provide adequate systems for standards development, assessment, and accountability. Performance *objectives* were clearly articulated; for example, by stating that the purpose of the act is to "safeguard people from possible injury, illness, or loss of amenity in the course of the use of any building."[12] But performance *standards* were insufficient, especially in the area of weathertightness.[13] The systems for assessment and accountability suffered from three major failings. First, inadequate requirements for inspections during building construction resulted in very little prospective government oversight: the government did not get involved until problems became apparent. Second, as builders innovated and used new materials and methods, local authorities and private certifiers had inadequate experience, guidance, and resources to predict the performance of such materials and methods, and to assess whether they would meet standards. Third, inconsistent assessment methods led to "jurisdiction shopping," with building activity concentrated in areas with the least demanding assessment procedures.[14]

New Zealand has not abandoned performance-based building codes, but regulators have amended the rules to allow for greater oversight from the central government, to improve the definitions of performance-standards, and to require more building inspections.[15] New Zealand's experience in particular highlights the importance of standard setting and assessment, and points to potential problems associated with regulatory flexibility that is not backed by accountability.

5.3 ADAPTIVE REGULATION

Adaptive regulation entails a planned process of periodic review and revision, used to bring new information and experience into prescriptive or performance-based regulations. Adaptive regulation is particularly applicable in fields that exhibit significant scientific uncertainty and the potential for discovery and change. The ideal procedure uses peer-reviewed scientific data to justify proposed changes, and subjects the policy to review at pre-planned intervals. Between the revisions, new information is collected, shortcomings of the existing regulations are exposed, and the technology and economy respond to the previous set of revisions. Over time, this process should create a regulatory structure that can better limit and deal with uncertainty and keep pace with evolving scientific understanding and new technology. Policy changes adopted as a result of a periodic review would not be retroactive (unless all affected parties agree) and, therefore, would typically not affect the compliance obligations of ongoing projects. They would, however, alter the regulatory environment for future projects.

In this section, our focus is on adaptive regulation at the program level. Adaptive principles can enhance regulation at the project level as well. For example, this could be done through regular review of monitoring plans to incorporate information from actual site performance. This type of project-level adaptability is best accomplished through use of performance-based regulation, as described above. Essentially, the performance standards would not change over the life of a permitted project, but the methods to achieve them could be adapted based on site performance and technologic advancements.

Successful adaptive regulation generally includes the following four components:

- a specific entity charged with the task of learning—of gathering evidence about the performance of current regulations and new knowledge relevant to the regulated activity;
- a specific entity with the authority to periodically make substantive changes to existing rules (this may or may not be the entity charged with learning);

- a timetable for periodic review and revision; and
- resources (including political will) to accomplish the review and the revision.[16]

The idea that federal agencies should use new evidence to improve old regulatory approaches (as opposed to just specific targets or timetables) has a long history in the United States. Plans were made to review existing regulations starting with the Administrative Procedures Act of 1946, and specific new plans made in the Administrations of Presidents Carter, Reagan, George H.W. Bush, and Clinton.[17] Most of these efforts yielded few results. Only eleven of the forty programs identified as candidates for regulatory revisions by the Bush Administration successfully revised their policies.[18] Clearly the idea of updating old regulatory approaches to incorporate new evidence has been more appealing in theory than in practice.

There are a number of possible reasons for the difficulty of revisiting established regulatory approaches. It is a demanding process that requires buy-in and cooperation from all parties involved. The agency that administers the regulations needs to willingly participate in the review of the policies. Agencies are often reluctant to undertake such an expensive and time-consuming task, especially if the regulated industry or special interest groups oppose the review in order to preserve aspects of the regulations that are advantageous to their interests. For instance, a recent review of the EPA's track record of updating standards under the Clean Water Act found that 72% of the 1,122 standards had never been revised.[19] However, despite the difficulties, adaptive regulation has been used at the federal level for some high-profile policies, including parts of the Clean Air Act and the Clean Water Act.

5.4 EXAMPLES OF ADAPTIVE REGULATION

A recent survey indicates that while feasible, adaptive regulation is still relatively rare in American rulemaking,[20] successful examples of adaptive regulation share certain traits; for example, when dealing with scientific issues, federal agencies have often worked with their own independent advisory groups or with the National Research Council to ensure that scientific experts are involved and that neither the agency nor the regulated industry wields undue influence over the process.[21] Additionally, the process of adaptive regulation is usually imposed by Congress, rather than initiated by agencies. Examples of programs that have used adaptive regulation are the Food and Drug Administration's (FDA) Post-Marketing Drug Surveillance Program, and the National Ambient Air Quality Standards (NAAQS) under the Clean Air Act (CAA). These programs designate entities to review and revise standards on a set timetable and devote resources toward completing the reviews. However, only the NAAQS program has resulted in substantive changes to previous regulations.

Post-Marketing Drug Surveillance Program

The FDA's Post-Marketing Drug Surveillance Program[22] monitors adverse effects of approved drugs in order to withdraw those that are later found to be unsafe based on patient outcomes.[23] The manufacturer of the drug is charged with the task of learning, and the FDA is charged with the task of revision.

The Post-Marketing Drug Surveillance Program has had mediocre results.[24] As discussed above, one of the components of a successful adaptive regulation program is resources, including motivation, to accomplish the review and revision. In this case, the manufacturer, the entity tasked with learning, has a disincentive to review and report adverse drug information because if adverse effects are found, the FDA could withdraw approval of the drug. This example illustrates the importance of designing adaptive regulation so that the entities responsible for review and revision have incentives to make the adaptive process successful.

National Ambient Air Quality Standards

The NAAQS, established by the Clean Air Act, are an example of a regulatory program that is both performance based *and* adaptive. The CAA requires the EPA to identify and control pollutants that "cause or contribute to air pollution which may reasonably be anticipated to endanger public health or welfare."[25] These pollutants are referred to as "criteria pollutants." While the basic regulatory structure does not change, for each criteria pollutant, the EPA must issue a performance standard (the NAAQS), based on an independent review of the latest scientific knowledge,[26] and must revisit this standard every five years, revising it as necessary to meet the public health and welfare standard.[27] The entity charged with the task of reviewing the NAAQS and proposing revisions is a seven-member "independent scientific review committee" appointed by the EPA, called the Clean Air Scientific Advisory Committee (CASAC).[28]

Experience with adaptive regulation under the NAAQS is mixed. Despite the mandatory schedule, the process has often gotten bogged down, and the courts have had to step in on multiple occasions to force the review. The EPA has set NAAQS for six criteria pollutants: Carbon Monoxide, Lead, Nitrogen Oxides, Sulfur Dioxide, Ozone, and Particulate Matter. Over the CAA's thirty-year history, the CASAC and the EPA should have conducted thirty reviews of the NAAQS. However, only twelve reviews have been undertaken (half of which resulted from court orders) and the EPA has adjusted the NAAQS—making them more or less strict—on only six occasions.[29]

However, one area where regular review and revision of NAAQS has had notable success is in reducing the harm caused by particulates. After a report showed that controlling particulate pollution resulted in 90% of the benefits from the CAA, the EPA funded an aggressive effort to advance relevant science to inform the periodic review.[30] The new research resulted in rule revisions that shifted the regulatory focus from visible smoke to fine particulates, which are now understood to have serious health implications.[31] The adaptive process played a significant role, both in the production of new knowledge and in its translation into regulation.[32]

5.5 ASSESSMENT OF CURRENT RULES FOR GEOLOGIC SEQUESTRATION

The positioning of the current GS regulatory framework on the continuum from prescriptive to performance based is largely a function of the new EPA UIC Class VI rules and the new mandatory GHG reporting requirements, both finalized in December 2010.[33] The UIC Class VI rules take an approach that includes both prescriptive and performance-based elements. Regulations to set minimum criteria for siting and for determining the area of review (the area that may be affected by GS) are performance based.[34] For example, a site is "suitable" for GS if owners can demonstrate that it has an injection zone that will receive the amount of CO_2 the project plans to inject, and a confining zone that will contain the CO_2.[35] No specific geologic characteristics are required. Rather, the onus is on the project developer to evaluate the suitability of a potential site using the testing and modeling methods he or she judges to be most appropriate. The regulator then has the responsibility to assess the quality and reliability of the developer's projections about how well the site will perform.

Other aspects of the Class VI rules are prescriptive, such as the sections governing well construction and operating requirements.[36] For example, the rule requires that every Class VI well be built with a "long-string casing" that extends from the surface to the injection zone and is cemented along its whole length.[37] A few aspects of the rule fall in the middle of the spectrum, containing both performance-based and prescriptive elements. The most important example is the section on monitoring requirements,[38] which prescribes specific procedures for monitoring factors such as injection pressure, corrosion, and mechanical integrity, but takes a largely performance-based approach to monitoring the extent of the carbon dioxide plume.

The GHG reporting requirements for GS sites are for the most part prescriptive, as is expected since their purpose is to create standardized, fully comparable measurements. Interestingly, however, they take a performance-based approach to monitoring and measuring one type of potential emissions from a GS site—emissions that would occur if CO_2 leaked from the injection formation, traveled all the way to the surface, and escaped into the atmosphere (surface leakage). Regarding surface leakage, the rule requires operators to submit a monitoring, reporting, and verification (MRV) plan which includes a "strategy for detecting and quantifying any surface leakage of CO_2."[39] In the preamble to this rule, the EPA notes that this performance-based approach allows for site-specific flexibility. However, the EPA also reported that they received many comments asking for greater clarity on how MRV plans will be evaluated.[40] The EPA addressed this concern by issuing a technical support document which discusses monitoring methods, potential strategies for detecting and quantifying surface leakage, sampling considerations, and detection capability of monitoring systems.[41] This technical support document does not, however, set standards for the expected level of performance. Rather, this responsibility is placed on the GS site operator, who is required to specify the detection limit (the targeted leak rate the proposed MRV plan could detect).

The structure that emerges from the combination of these regulations is that the regulatory framework for GS established by the EPA (through the UIC Class VI rule and the GHG reporting rule for GS sites) tends toward being prescriptive for engineered components of the GS system, and performance based for geologic components of the GS system. One way of viewing this is that performance-based approaches have been chosen for the aspects of GS project operation about which technical and scientific uncertainty is greatest. This has the potential to promote learning and creativity, but carries certain risks. For one, the performance objectives are not supported by clear standards, and monitoring procedures are not yet well tested. While the lack of standards in the performance-based aspects of the current GS regulations is understandable, as data to derive them do not yet exist, this deficit should be rectified as experience is gained. Through periodic revisions (as described in the next section), regulators should use the expanding GS knowledge base to establish standards for the desired level of performance at GS sites, and to identify assessment procedures that can be used to establish that the project is complying with permit requirements.

There is no mandate in the present EPA regulatory approach for program-level review and revision of GS regulations. The UIC Class VI rules include a number of elements that promote adaptability at the project level; for example, operators must reevaluate the area of review,[42] and review the site testing and monitoring plan[43] at least every five years. Project-level regulatory adaptations will provide valuable information for program-level adaptations, and the EPA acknowledges the importance of an adaptive approach for the UIC program in the preamble to the Class VI rule, stating that they plan to review the rulemaking every six years to determine whether modifications to the Class VI requirements are necessary.[44] However, despite the fact that the EPA has the authority to revise rules, experience shows that revisions are rare unless they are legally mandated.[45]

5.6 RECOMMENDATIONS ON LEARNING AND ADAPTATION

We recommend the creation of a performance-based, adaptive regulatory framework that both recognizes the uncertainty associated with GS and seeks to systematically decrease it over time. By emphasizing performance-based rules to encourage flexibility and learning, and by mandating a systematic process to regularly review and improve the rules, performance-based adaptive GS regulation should ensure that risks to health, safety, the environment, and property are well managed, that compliance costs are minimized, and that investors have the regulatory certainty required to make the essential large investments in GS projects. Our specific recommendations, detailed in the model legislation reproduced in the Appendix, are as follows:

Congress should enact new GS legislation that mandates periodic review and revision of the regulatory framework for geologic sequestration. History shows that the EPA's regulatory authority to revise existing rules needs to be supplemented by a legal mandate to do so. Much will be learned from the operation of early GS projects, and this knowledge should be used to improve the existing GS regulatory framework. The revised rules should apply to all new GS projects. In order to provide regulatory certainty, regulatory changes should not apply retroactively to existing projects unless all affected parties agree.

New legislation should create an independent technical advisory committee that is charged with learning from experience and advising the EPA and the new Federal Geologic Sequestration Board (FGSB) that we propose in Chapter 7, as well as other agencies, on ways to improve the regulatory framework for GS. The EPA should contract with the National Research Council (NRC) to establish an independent GS Technical Advisory Committee (GSTAC) that will meet at least once every seven years to advise the EPA, the proposed FGSB, and other agencies involved in the regulation of GS, on ways to improve the regulations. The objectives of the GSTAC are to use field experience from GS sites in the United States and elsewhere, including data and measured performance at specific sites, to inform the ongoing management of risk and to provide a means to improve GS regulation. Each time the GSTAC meets, it should issue a full report that includes recommendations to Congress and the agencies on whether and how the GS regulatory framework should be changed to reflect the latest experiential learning from the field. These changes include:

- modifications to performance-based elements of the regulations, such as clearer articulation of objectives, specification of performance standards, guidance on assessment methodologies to insure that performance standards are met, and requirements for overall accountability;
- changes to prescriptive elements of the regulations to make the requirements reflect the best current science, or to move them toward being more performance based;
- adjustments in how the regulatory authority for GS is balanced between EPA programs (e.g., UIC program, GHG reporting program, CAA programs), and between federal and state regulators, concentrating on streamlining oversight, reducing duplication, and eliminating regulatory gaps;
- new legislation, or amendments to existing laws that could improve the safety, environmental performance, economic efficiency, and/or GHG reduction capacity of GS;
- other changes as needed based on scientific, technologic, and economic developments.

The GSTAC reports would provide a state-of-the-art summary for GS project developers, regulators, and financial institutions. Although we recommend that the GSTAC be established under the NRC, the review could alternatively be conducted by a special committee of the EPA Scientific Advisory Board or by a stand-alone advisory committee created by Congress. Using the NRC is preferable because that would better assure the high level of independent expertise that is necessary to recommend comprehensive, coordinated, performance-based GS regulation under the UIC program, the GHG reporting program, and any other applicable regulations.

Results from early projects should be as transparent and as publicly available as possible. This will enhance learning and decrease uncertainty. In particular, the EPA should compile and report methods and results of site characterization and monitoring studies in a format that is easily accessible to state and federal regulators, GS project developers, and academics. All monitoring results from GS sites, including UIC Class VI permit reporting (collected by EPA Office of Water) and GHG reporting (collected by EPA Office of Air) should be consolidated for ease of use.

Congress should create a mechanism that provides adequate funding to support GS regulators. Regulators will need significant support and training in order to evaluate GS site permit applications and proposed monitoring plans. They should have ample opportunity to review the experience of GS sites in other parts of the country and to consult with other regulators regarding performance levels of GS sites. This funding, which we propose should come as part of a per-tonne injection fee paid by project operators (see Chapters 4 and 7), will promote regulatory consistency and industry accountability.

6

ACCESS TO PORE SPACE FOR GEOLOGIC SEQUESTRATION[1]

To perform geologic sequestration, carbon dioxide gas is compressed to a supercritical fluid and injected approximately a kilometer or more below the surface into the microscopic pore space of rock matrixes. Injected CO_2 flows through and fills a portion of the pore spaces in permeable layers of the rock matrix, displacing any other fluid (typically briny water) that might be there, as shown in Figure 6.1. In a well-designed GS project, upward migration of the CO_2 is prevented by less permeable overlying layers of "cap rock."

Depending on volume injected, the formation geology, and the depth, porosity, and permeability of the injection zone, sequestered CO_2 from a single project could spread over large areas (many hundreds of square kilometers in the case of thin formations) and subsurface pressure effects can extend to even larger areas. Growing interest in developing commercial-scale geologic CO_2 sequestration (GS) has sparked an intense debate about the ownership and commercial use of pore space. In fact, the commercial viability of CO_2 sequestration may, in large part, depend on whether and how property rights issues are resolved.[2] Prior to injecting CO_2 into the subsurface for permanent geologic sequestration, the injecting party must either own the pore space, have permission from the owner, or have statutory or common law right to use the pore space that avoids potential liability or exposure to trespass and nuisance claims. Clearly, before a geologic CO_2 sequestration field can be developed, the project developer will want to acquire the authorization to access and use pore space to avoid liability for subsurface trespass and nuisance. If a GS project development receives authorization from a state or the federal government to use pore space and the authorization shields the developer from trespass to any degree, "takings" law will certainly be implicated as well. This chapter considers the legal and commercial models for securing the rights to use geologic pore space in an effort to sequester billions of metric tons of CO_2 deep underground to mitigate climate change.

FIGURE 6.1 *Pore space is blue and grains of quartz are white in this photograph of a microscopic cross-section of rock. (Figure courtesy of CO2CRC.)*

6.1 COMPETING USES OF THE SUBSURFACE

Because of the potentially large size of commercial geologic sequestration projects, other economic uses of the subsurface, such as hydrocarbon production, natural gas storage, fluid waste disposal, and groundwater recovery and storage activities may coexist or compete with subsurface CO_2 injection.[3] It is likely that CO_2 sequestration projects will overlap these other subsurface enterprises. At least in early decades this will especially be true of enhanced oil and natural gas recovery projects that inject CO_2 as a method for repressurizing production fields. State legislatures, particularly in oil and gas producing states, are already enacting GS-specific legislation to avoid conflict between different economic uses of the subsurface.[4]

Many oil and natural gas production wells operate at average depths similar to those of proposed CO_2 sequestration projects, roughly one kilometer or deeper.[5] Subsurface formations with hydrocarbon-bearing strata are typically well-characterized and are often stacked between non-hydrocarbon-bearing saline aquifers.[6] Thus the possibility of developing a CO_2 sequestration site above or below oil or natural gas reservoirs or within depleted reservoirs may have the advantage of reducing characterization and capital costs compared to an uncharacterized site, but doing so could also create potential interference between projects.[7] For instance, in a reservoir producing natural gas, migrating CO_2 could commingle directly with the gas, both reducing the success of the sequestration and requiring efforts to remove the CO_2 from the gas production stream.[8] At the same time, soluble CO_2 could cause the precipitation of carbonate minerals and plug flow paths, which would reduce the extraction efficiency for existing hydrocarbon production facilities.[9] The pressure effects from the injection operation, particularly if multiple sites are used to inject CO_2 into a single basin, could also adversely affect other injection operations by potentially altering injectability, plume size and shape, and associated monitoring.[10]

Hydrocarbon production also produces large amounts of wastewater—an average of seven gallons of water is produced for each gallon of oil. The produced water must be separated and disposed of safely, usually by underground injection. Over 750 billion gallons of oil-produced waters are injected into the subsurface through 150,000 disposal wells in the United States each year.[11] This volume of produced water is the rough equivalent of the volume that two gigatonnes (Gt) of CO_2 would occupy at a depth of one kilometer.[12] The wastewater is generally handled on site, with approximately one-quarter of it being injected back into the oil-producing formation, in part to increase oil production.[13] Other operators inject the produced waters into non-producing formations at varying depths where formations of adequate porosity and permeability are present.[14] Some wastewater disposal wells inject below the hydrocarbon formation and others inject above it.[15] In Texas, produced water is injected into non-producing formations varying in depth from 300 to 3,000 meters, with 60% of these wells being a kilometer or more deep.[16] While the level of regulatory oversight under UIC rules is much less demanding for these waste waters than those that will be imposed on CO_2 by Class VI rules, both the practice and the scale of handling produced water is similar to those expected for permanent geologic sequestration of CO_2.[17]

Underground natural gas storage is another area where use of the subsurface for CO_2 sequestration may require coordination. Today, in the United States, there are roughly 130 natural gas operators storing between 1,200 and 3,300 billion cubic feet (BCF) of natural gas throughout more than 400 active storage facilities in the lower-48 states.[18] Gas storage has helped to balance fluctuations in supply and demand of natural gas for nearly 100 years. Similar to CO_2 sequestration, depleted hydrocarbon fields and saline aquifers are commonly used for natural gas storage.[19] If care is not exercised in reservoir design, CO_2 and stored natural gas might commingle. This would both reduce the effectiveness of sequestration and degrade the quality of the stored natural gas.[20]

In addition, long-standing and new uses of the subsurface for activities wholly unrelated to hydrocarbon production may take place in formations and depths similar to CO_2 sequestration. For example, hazardous and non-hazardous fluid wastes and municipal wastewater are often disposed of below the lowest underground source of drinking water by injecting them deep into subsurface formations.[21] As outlined in Chapter 4, the Environmental Protection Agency and delegated state agencies have regulated the underground injection of fluid wastes under the Underground Injection Control program by creating "classes" of injection wells and setting standards for injection to protect underground sources of drinking water. These waste injection wells are located in formations where freshwater is protected from the injection zone by an impermeable cap rock or confining layer, much like what would be used for CO_2 sequestration. Injection zones typically range from slightly over 500 meters to more than 3,000 meters in depth.[22] There are roughly 550 Class I waste wells in the United States, mostly located in the sedimentary basins of the Gulf Coast and Great Lakes regions.[23] While approximately 48% of the Class I wells are for non-hazardous liquids, another 30% of the wells are dedicated to municipal wastewater disposal in Florida, where over 3 billion metric tons of wastewater are injected annually.[24] Most of these wells are operated with little or no approval from the owners of surface properties under which the injection occurs, except for the specific surface parcel on which the injection well is located.

Finally, compressed air energy storage (CAES) and underground aquifer storage and recovery of groundwater (ASR) are becoming increasingly attractive possible uses of the subsurface. CAES could help manage complications imposed by the intermittency of large-scale electricity produced by wind. Surplus electricity produced by wind that would otherwise flow into the electric grid could instead be used to compress air that is pumped and stored in deep geologic reservoirs to be used later to make natural gas turbines operate more efficiently. Given the rather different geology that is suitable for CAES, it is unlikely that conflicts will arise with CCS projects. ASR involves injecting water into deep underground reservoirs for later retrieval. A handful of states have mature permitting regimes to facilitate the storage of freshwater underground so that it may be withdrawn during dry periods. ASR is thought to be a promising solution for the future of freshwater management.

As the preceding paragraphs suggest, there is a real potential for GS operations to interfere with actual or reasonably foreseeable uses of subsurface pore space. Currently, there is little to no federal or state statutory authority that governs issues of subsurface property rights in the context of CO_2 sequestration. Moreover, many subsurface injection activities discussed above are permitted and regulated by different federal and state agencies. The federal Safe Drinking Water Act gives the EPA authority to manage the UIC program, which regulates underground fluid waste injection activities and enhanced oil recovery, but not natural gas storage.[25] Many states have asserted primacy to administer the UIC program, but do so through more than a single state agency. For example, in many petroleum-producing states, oil and gas conservation agencies manage the UIC program for saltwater disposal and injections related to petroleum operations, while another agency manages other UIC matters.

As explained in Chapter 4, in November of 2010, the EPA published its final rules for managing the injection of CO_2 for geologic sequestration under the UIC program.[26] The rules include provisions for on-site characterization, well construction and operation, post-injection monitoring, and post-closure stewardship.[27] The EPA stated, however, that the SDWA does not give the agency any authority to address GS-specific property rights concerns; therefore, these issues are not addressed in the new rule.[28]

6.2 WHO OWNS PORE SPACE IN THE US?

In the United States, property rights historically have been defined by state law rather than federal law. Common law property rights are generally viewed as a bundle of rights that together define how ownership of various resources are divided, and establish limitations on the use of each by the owner. A typical bundle of rights might be: 1) surface rights; 2) rights to coal; 3) rights to natural gas; 4) rights to oil; 5) rights to minerals other than coal, oil, and natural gas (sometimes referred to as residual mineral ownership); 6) rights to groundwater; and 7) rights to engage in storage (e.g., natural gas and freshwater). These various interests may be burdened by licenses, leases, servitudes, and security interests.

Today, there exists no uniformity in the way that the right to inject fluids of any kind into deep subsurface pore space is legally recognized, acquired, or authorized. In the United States, under the common law maxim *cujus est solum, ejus est usque ad coelum et ad inferos* (commonly known as the "*ad coelum* doctrine"),[29] a fee simple[30] owner of land holds title to the entire tract from the heavens to the depths of the earth. Under this maxim, a fee simple owner would own the subsurface pore space. The question of pore-space ownership most commonly arises when the fee simple interest is severed into a surface estate and one or more separate mineral interests. As between the surface owner and mineral owner, few states have statutorily or judicially determined who owns the pore space. In many states, the questions of pore-space ownership and use rights would be ones of first impression. To date, only a handful of cases across the country have addressed this issue, with the majority and the better-reasoned decisions holding (given the specific facts before the court) that the surface owner owns the pore space.[31] However, even where a surface owner may "own" pore spaces, the mineral owner has an implicit right of reasonable use to facilitate enjoyment of mineral rights. Because mineral rights may be severed by specific substances or by depth, are frequently fractionalized among multiple parties, and are subject to a multitude of transactions and property interests, a single tract of land could be subject to a variety of mineral claims held by literally dozens—perhaps hundreds—of claimants.[32]

The emerging opinion that pore-space rights generally belong to the surface owner is not founded on the idea that surface rights and pore-space rights are inextricably bound together.[33] Instead, the opinion is based on the recognition that, as a historical matter: 1) owners of fee simple estates traditionally have owned everything on, above, or below the surface except to the extent particular rights have been granted to others; 2) historically, fee simple owners have tended to retain ownership of the surface when carving mineral interests out of the fee for transfer to others; 3) legal instruments transferring a portion of the fee simple owner's property rights to others have tended to be narrowly drafted; and 4) courts generally interpret such instruments to have transferred only what is specifically mentioned together with whatever other rights are necessarily associated with the rights explicitly identified.[34] For example, legal documents creating mineral interests have typically used narrow language such as "oil, gas and other minerals" rather than broad language that would sever everything below the surface of the land and transfer it to a new owner. In the former case, the particular mineral interest will generally have the right to use the pore space as reasonably necessary to extract minerals, but is not likely to "own" the pore space or to have the right to use the pore space for purposes unrelated to extracting minerals.[35] The primary exception will be where the original fee simple owner sells surface rights but expressly reserves all subsurface rights—a rare practice. Thus in most cases, American jurisdictions, grounded in English Common Law, will most likely find that pore space is owned by surface owners.

Several states have begun to develop regulatory frameworks to manage geologic sequestration of CO_2, with specific attention directed toward the issue of pore-space ownership. Wyoming, Montana, and North Dakota passed legislation explicitly defining pore-space ownership.[36] Wyoming H.B. 89 addressed the issue of property rights by stating that "[t]he ownership of all pore space in all strata below the surface lands and waters of this state is declared to be vested in the several owners of the surface above the strata."[37] In 2009, the Wyoming governor signed into law H.B. 57, which amends the pore-space provision in H.B. 89 and clarifies that the mineral estate is dominant over the surface estate.[38] That same year, North Dakota S.B. 2139 similarly proclaimed that "[t]itle to pore space in all strata underlying the surface lands and waters vested in the owner of the overlying surface estate."[39] North Dakota's bill further attaches pore-space rights to the surface estate by prohibiting severance of pore space from the title to the overlying surface property.[40] Montana S.B. 498 creates a presumption that the surface owner owns subsurface pore space if deeds or other severance documents do not demonstrate otherwise.[41] Like Wyoming and North Dakota, Montana's new statute explicitly does not interfere with common law or the dominance of the mineral estate. In 2011, West Virginia's CCS working group recommended that legislation declare that pore space within the boundaries of the state and (a) 2,500 feet beneath the surface or (b) between 2,500 feet and 12,000 feet beneath the surface, that are not under an existing or reasonably forseeable use by the respective property owner, are part of the public domain.[42]

6.3 DOES THE USE OF PORE SPACE FOR GS REQUIRE COMPENSATION UNDER THE LAW?

The Takings Clause of the Fifth Amendment to the US Constitution instructs that private property shall not "be taken for public use, without just compensation."[43] The Supreme Court of the United States explained that the Takings Clause "was designed to bar Government from forcing some people alone to bear public burdens which, in all fairness and justice, should be borne by the public as a whole."[44] The Constitution does not create or define the scope of property interests that are protected under the Fifth Amendment, but instead requires compensation in the event that an impairment of those property interests amounts to a taking.[45] To determine whether a protectable property interest exists, courts look to "existing rules or understandings" and "background principles" derived from sources such as Federal and state common law.[46] Even though property rights typically fall under the purview of state law, state-created property rights may be limited by federal law.[47]

In the context of CO_2 sequestration, the threshold question is therefore whether a surface owner or mineral owner has sufficient interests in subsurface pore space to implicate the Takings Clause of the Fifth Amendment to the Constitution, or similar limitations imposed by state constitutions. There is little dispute that, subject to reasonable regulation, property owners have significant rights to use their property as they see fit. Just as importantly, property owners have the right to exclude others from making use of their property without consent. If the federal or state governments wish to confiscate private property, condemn private property, or authorize third parties to confiscate or condemn private property for a public purpose, the government may do so through the exercise of eminent domain authority. Eminent domain authority is conditional, however, and requires that "just compensation" be paid to the property owner.

If a government action is challenged as having impaired an established property interest, a court must decide whether to analyze the action as a physical taking or as a regulatory taking.[48] A physical taking occurs when the government engages in, or authorizes a third party to engage in, a permanent physical occupation of private property.[49] In the case of physical invasions, "no matter how minute the intrusion, and no matter how weighty the public purpose behind it," a taking has occurred and just compensation is required.[50] Even when there is no physical occupation of private property, a regulatory taking can occur if government regulation places too great a burden on the owner's use of the property.[51] A regulatory taking can take place under two circumstances.[52] First, a regulatory action can entail a per se taking when the regulation completely deprives a property owner of *all* economically beneficial use of the property.[53] Second, in the absence of a complete deprivation of all economic use of property, courts will consider whether the regulatory restriction rises to the level of a compensable taking under the multifactor balancing test prescribed in *Penn Central Transportation Co. v. New York City*.[54] The *Penn Central* balancing test, as it is commonly known, considers: 1) the character of the government action; 2) the severity of the economic impact; and 3) the extent to which the regulation interferes with the property owner's distinct, "investment-backed" expectations.[55] However, even if a government action constitutes a physical taking or regulatory taking, a violation of the Fifth Amendment occurs only if the taking is without "just compensation."[56] Consequently, if a court concludes that the monetary value of property owner's net loss as a result of the taking is zero, the compensation due under the Constitution is also zero.[57]

The questions of how far up into the sky and down into the earth property rights extend and to what extent these rights are protected are also important. As noted above, the *ad coelum* doctrine instructs that the rights of the surface owner extend up to the heavens (*ad coelum*) and down to the center of the earth (*ad infernos*). However, since the advent of air travel in the early part of the twentieth century, the doctrine no longer applies in absolute terms to ownership of airspace high above the ground. The opinion that ownership of airspace rights "extended to the periphery of the universe" was laid to rest at the national level by the Supreme Court in 1946 in the case of *United States v. Causby*.[58] *Causby* established a precedent that allows trespass and takings claims related to airspace use to prevail only in those circumstances where there was "a direct and immediate interference with the enjoyment and use of land."[59] In reaching its decision, the *Causby* court stated that the *ad coelum* doctrine "has no place in the modern world."[60]

Courts have looked to *Causby* and other airspace cases when deciding subsurface property rights cases; however, case law involving subsurface property rights is much broader and more complicated than the body of airspace case law. The complication primarily arises from several attributes of subsurface rights that fundamentally distinguish them from airspace rights.[61] The first distinction is that, unlike airspace rights, subsurface rights have been severed, conveyed, bought, sold, used, and developed by private parties and federal, state, and local governments since at least the mid-1800s, and perhaps before. This has resulted in ownership, use, and exploitation of the subsurface in a manner far more diverse and tangible than ever existed for airspace rights. A second distinction is that the airspace cases addressed a single and very compelling "public interest"—national air travel—competing against surface interests. By contrast, in cases involving subsurface rights, the surface owner's rights often clash with multiple, but more local, competing uses. Moreover, these competing uses are subject to federal or state regulatory systems designed to promote each activity in the public interest. Thus, there is reason to believe that, at least at the depths of several kilometers, the judicial and legislative precedent which limited the protection of private property rights in airspace may not be entirely instructive in the context of subsurface property rights in general and subsurface pore-space rights in particular.

6.4 ALTERNATIVE MODELS FOR THE ACQUISITION OF THE RIGHT TO USE PORE SPACE FOR FLUID INJECTION

Modern case law generally treats surface intrusions and subsurface intrusions differently. Regarding traditional surface trespass, *Loretto v. Teleprompter Manhattan CATV Corp.* is perhaps the seminal case in which the Supreme Court addressed the issue of permanent physical occupation of a surface owner's property.[62] The specific issue before the court was whether a New York law requiring a landlord to permit the installation of a cable company's cables on rental properties to furnish cable television services to tenants rises to the level of a taking without just compensation. The Supreme Court ruled that the state statute amounted to a taking of a portion of the plaintiff's property—around one and one-half cubic feet on the outside of the rental building, to be precise—for which she was entitled to just compensation under the Fifth Amendment.[63] The court found that, to the extent the government permanently occupies property or grants a third party the right to do so, it effectively destroys the right to "possess, use, and dispose" of property and will amount to a taking in any such circumstance.[64] The court further concluded that, in applying this per se taking rule for physical occupations, the size of the area occupied is irrelevant, as is whether the plaintiff previously occupied the space in question.[65] On remand, however, the Court of Appeals of New York[66] ruled that the amount of compensation awarded could be nominal and predetermined (in *Loretto*, a one-time $1 payment), provided that property owners had a mechanism available through which to seek more compensation by establishing that the diminution in value of the property was materially different from the general assumption, and this therefore entitled the owners to receive greater compensation.

While no case specifically related to the encroachment of subsurface property rights has yet been argued before the Supreme Court, state and other federal courts have ruled on this issue in the context of various commercial and industrial subsurface injection activities that gave rise to both takings and trespass claims. These courts have a history of balancing competing interests in the subsurface and have placed great weight on the public interest and regulatory approval associated with certain activities. Specifically, courts have given surface and mineral owners only limited protection against subsurface intrusions resulting from the migration of fluids injected underground.

Courts have addressed the issue of subsurface trespass in the context of five subsurface injection activities that are frequently considered analogous to geologic sequestration of CO_2: 1) licensed underground natural gas storage projects; 2) licensed underground fluid waste injection and disposal projects; 3) state-authorized unitization for enhanced petroleum recovery; 4) hydraulic fracturing for petroleum production; and 5) underground freshwater storage and recharge. In each of these groups of cases, courts balanced the need to protect a landowner's right to exclusive possession and enjoyment with the public-interest need of facilitating valuable enterprises that meet important societal needs.[67] Although not large in volume, the weight of case law supports the proposition that takings and trespass claims will not stand absent actual and substantial damages. However, the case law is neither entirely unified nor coherent.

In the cases involving underground fluid waste injection, enhanced oil recovery, hydraulic fracturing, and underground water storage, the courts have consistently modified the *ad coelum* doctrine and limited the ability of surface owners and mineral owners to recover money damages for trespass or for a taking, resulting from government-authorized use of subsurface pore space. Regarding subsurface natural gas storage, project developers have generally proceeded under the Natural

Gas Act to acquire subsurface property by eminent domain, thus implicitly acknowledging (or at least not expressly challenging) that the use of pore space requires compensation; however, where eminent domain was not utilized, no natural gas storage case has awarded damages for trespass. Courts and lawmakers will undoubtedly look to these cases for guidance to determine whether the use of deep geologic pore space for permanent sequestration of CO_2 without compensation rises to the level of a trespass or unlawful taking.

The Natural Gas Storage Model

While the natural gas storage case law is instructive of how courts may treat the use of subsurface pore space in the context of geologic CO_2 sequestration, the number of opinions available for review is sparse. What is more, most judicial decisions that address subsurface invasions caused by the storage of natural gas either focus on ownership of the stored gas, not the invasion itself, or treat trespass allegations as actions in inverse condemnation[68] because gas storage rights may be acquired by eminent domain.[69] Thus, any attempt to distill from this body of jurisprudence definitive legal principles that might apply to the use of pore space in the context of geologic CO_2 sequestration is, at best, of limited utility. At its worst, conclusions drawn from such an analysis could lead, and perhaps have led, lawmakers dangerously astray. Arguably, the practice of compensating property owners to use subsurface pore space for storing natural gas developed as much out of industry custom and the need to maintain possession and control of injected gas as it has from adherence to legal and regulatory requirements. Because natural gas is a valuable commodity that is stored only temporarily and then retrieved, a gas storage operator has a compelling motivation to compensate landowners for use of the storage space to ensure complete control of the storage reservoir and to protect against losing gas before it can be recovered and sold or consumed. By contrast, GS operators will not be concerned with maintaining the integrity of the sequestration field so that it can be easily recovered at a later time to be traded and sold. Quite the contrary, the sole purpose of GS is to keep CO_2 underground and out of the atmosphere permanently. Thus, apart from assuring the long-term integrity of the storage reservoir, the migration of CO_2 will be of no concern to the GS storage operator.

To facilitate the development of natural gas storage projects, many states adopted condemnation statutes, which allow gas storage companies to condemn rights in underground storage reservoirs. However, many of these states allow condemnation only in reservoirs where minerals can no longer be produced in commercial quantities.[70] When a gas storage facility is developed and is later found to interfere with active mineral production operations, most state condemnation statutes are not designed to remedy the problem.[71] Often the only solution is to discontinue storage operations. Of course, state condemnation statutes used to develop underground gas storage facilities do not currently address GS. Legislation, either state or federal, is needed to provide condemnation authority for GS.

Enhanced Hydrocarbon Recovery, Field Unitization, and Hydraulic Fracturing Models

For decades, oil and gas production companies have injected fluids (often water, natural gas, or CO_2) into producing oil and gas reservoirs to enhance the ultimate recovery of hydrocarbons. This process, known as enhanced oil recovery (EOR) or secondary recovery, repressurizes the reservoir and can significantly increase petroleum recovery. Trespass issues can arise when an EOR operator injects fluids into the subsurface of their own property and the fluid then invades the subsurface of neighboring property. Compensable damages can result when oil reserves on the invaded property are displaced or when the invasion makes recovery of such reserves more difficult or expensive. Judicial decisions addressing subsurface trespass allegations in the enhanced-recovery operations are mixed, but several cases suggest that a compensable trespass claim is less likely to succeed if a regulatory agency authorized the particular operation brought before the bench for trial.

Most enhanced-recovery operations take place in fields that have been "unitized" pursuant to state regulatory orders. With "field unitization," individual producing tracts are combined, with production being carried out by a designated unit operator, while the costs and profits of the project are shared by owners of the unitized tracts on a "fair share" basis. In most states, the property rights of landowners in the unitized area are addressed by state unitization laws, which may allow the unit operator to proceed regardless of whether they are able to reach agreement with all landowners. Most states require agreement by a minimum percentage (ranging from 50% to 85%) of ownership in a reservoir to authorize compulsory unitization.[72] On the other hand, in Texas, only voluntary unitization is possible, although the state regulatory agency, the Texas Railroad Commission, still issues orders approving voluntary units.[73] Even when unitization of the entire reservoir does not occur, regulatory officials may approve agreements creating a partially unitized field. In those circumstances, common law has evolved to protect unit operators against trespass claims made by "holdout" landowners who decline the opportunity to participate in the unit. This protection protects unit operators from trespass claims as well as from claims for excessive drainage of petroleum and is grounded in the public's interest in efficient petroleum production. Allowing holdouts within the unit area to sue for damages would defeat the state's goal of facilitating resource conservation through unitized production.

As with title issues, regulatory bodies have no general authority to authorize trespasses or other torts. However, an early Texas case suggests that regulatory orders may provide some protection. In *Railroad Commission v. Manziel*,[74] the court stated that it was presented with the issue of "whether a trespass is committed when secondary recovery waters from an authorized secondary recovery project cross lease lines."[75] After discussing the utility of EOR operations, the court stated,

> We conclude that if, in the valid exercise of its authority to prevent waste, protect correlative rights, or in the exercise of other powers within its

> jurisdiction, the Commission authorizes secondary recovery projects, a trespass does not occur when the injected, secondary recovery forces move across lease lines, and the operations are not subject to an injunction on that basis. The technical rules of trespass have no place in the consideration of the validity of the orders of the Commission.[76]

The court further noted that "To apply the general rules of surface invasions would interfere with the public policy considerations behind secondary recovery operations which, the court found, should be encouraged as a matter of 'public necessity.'"[77]

In all of the EOR cases—adjudicated not only in Texas but in many other states as well[78]—the courts have placed great emphasis on the states' statutory policies encouraging enhanced-recovery operations to promote the public's interest in efficient petroleum production. The courts focused on the existence of a state regulatory body to balance the needs of various rights-holders and refused tort recovery for those who declined to participate in unitization. These cases do, however, signal willingness by the courts to allow future plaintiffs recovery where there is actual damage to, or interference with the use of, a plaintiff's tangible property. Allowing recovery for actual damage to property is different from finding that a landowner possesses the type of property right in the subsurface that empowers the owner to prevent others from injecting fluids into the pore space underlying the owner's property; it is this type of absolute ownership doctrine the courts seem to have clearly rejected in the context of EOR. This judicial approach to pore-space rights supports the notion that GS project developers could be authorized to access and use pore space without being required to seek permission from landowners, or compensate landowners, for the right to permanently sequester CO_2 in the deep subsurface.

Underground Fluid Waste Disposal Model

An activity perhaps more closely analogous to CO_2 sequestration than natural gas storage or EOR is the underground injection of waste fluids. Unlike the natural gas storage case law, the cases on underground waste injection show that courts have consistently rejected any notion that property owners have the absolute right to prevent the underground migration of fluid waste into their pore space, or are entitled to compensation for such use of their pore space, when the waste injection is conducted pursuant to regulatory approval. And, unlike EOR, fluid waste disposal operations do not involve the unitization of the underground reservoir—neither costs nor profits of the disposal operation are shared by affected landowners. In each fluid waste case, the courts have placed great weight on the public interest and regulatory approval associated with the underground injection of fluid wastes, modifying the common law doctrines relating to subsurface property rights accordingly.[79] At the same time, though, each of the courts held open the possibility that a plaintiff could recover damages if it could show that the migration of injected waste caused actual harm to, or interference with the use of, its property.[80]

While there is no physical difference between the subsurface pore space used for underground waste injection and natural gas storage—in fact, both activities may be carried out in the same geologic formations (where the geology may also be suitable for geologic sequestration of CO_2)—the judicial protection afforded to subsurface property rights in the context of waste injection is quite different from for natural gas storage. In much of the nation, it appears that most underground waste injection operations conducted pursuant to federal or state authorization under the UIC program that do not cause actual harm to adjacent properties may be carried out without compensation being paid to the surrounding landowners because the activity is considered necessary and in the public interest. For example, municipalities in Florida appear to be injecting roughly 3 billion tonnes a year of treated wastewater without the consent of subsurface property owners.

Attempts by landowners to prevent or exclude subsurface waste disposal have not been successful in the absence of establishing actual and substantial harm to the use and enjoyment of their property. For example, in the landmark case *Chance v. BP Chemicals, Inc.*,[81] BP Chemicals secured a Class I underground injection well permit to dispose of hazardous chemical waste.[82] Subsequently, neighboring landowners initiated a class-action suit wherein they asserted that the injected waste trespassed into their subsurface pore space.[83] The *Chance* court ultimately concluded that a landowner's subsurface right to exclude others extends only to invasions that "actually interfere with the [landowner's] reasonable and foreseeable use of the subsurface."[84] The Ohio Supreme Court placed significant weight on the fact that the plaintiffs had no specific evidence that the defendant's wells were causing any problems, only opinion testimony that problems may arise in the future. In other words, a landowner may not recover damages for mere loss of speculative value. Since the injection of hazardous waste by BP Chemicals was not interfering with any reasonable and foreseeable use of the plaintiffs' property, the court held that migration of the waste into neighboring pore space was not compensable. The *Chance* court went on to declare, "Just as a property owner must accept some limitations on the ownership rights extending above the surface of the property, we find that there are also limitations on property owners' subsurface rights."[85] If *Chance* is followed, GS project developers would not need to acquire pore-space rights prior to commencing operations, only an injection permit. However, should injected CO_2 cause actual damages to neighboring properties or impair "reasonable or foreseeable" uses of pore space owned by neighboring property owners, the sequestration project operator would be liable for trespass damages, although the project itself would not be enjoined.

Groundwater Aquifer Storage and Recharge Model

A handful of states have mature permitting regimes to facilitate the storage of freshwater underground so that it may be withdrawn during dry periods. Aquifer storage and recovery (ASR), as this method of water storage is commonly known, is a promising solution for the future of freshwater management.[86] However, in spite of relatively well-developed permitting programs in several states, courts have rarely discussed the issue of subsurface property rights in the context of underground storage of fresh water.[87] California and Colorado are the exception. Both states have dealt with the issue directly, and both concluded that aquifer storage space is a public resource.[88]

The ASR cases highlight that courts often treat the protection of property rights somewhat like a moving target, one which is contingent upon and balanced against the promotion of important and well-defined public interests. If California's ASR approach to subsurface property rights is applied to GS, federal or state legislatures would declare that the permanent sequestration of CO_2 in geologic formations furthers an important public interest—reducing greenhouse gas emissions to the atmosphere—and impose, through a valid exercise of police power, a public servitude over deep geologic formations suitable for permanent sequestration of CO_2. If an ASR framework is applied, a GS project developer may not need to compensate property owners to use pore space for CO_2 sequestration. This is especially likely if no harm to or interference with the use of subsurface property is caused by the migration of injected CO_2.

6.5 POTENTIAL LEGAL FRAMEWORKS FOR MANAGING GS ACCESS TO PORE SPACE

At present, there is limited state-level authority and practically no federal-level authority for handling the subsurface property rights issues associated with the use of pore space for CO_2 sequestration as well as other commercial and industrial subsurface injection of fluids. While the Environmental Protection Agency has the authority under the Underground Injection Control program to permit and regulate geologic sequestration of CO_2, the agency currently has no authority to consider the subsurface property issues attendant to permanent geologic CO_2 sequestration. A legal framework for GS is needed that balances the interests of private property owners with the public benefit of sequestration, and reduces the possibility of interference with other commercial uses of the subsurface that are also in the public interest. To be workable, such a framework must allow regulators to consider the trade-offs between private interests and the public benefit of a proposed GS project and determine the most equitable and beneficial use of the pore space. A successful framework would increase the potential for either avoiding most subsurface property disputes outright or resolving them at the outset in a stable and predictable environment. The remainder of this chapter evaluates a range of common law, statutory, and regulatory approaches to managing the access and use of pore space for geologic sequestration of CO_2, and concludes by proffering a framework for addressing this issue in a manner that is fair and equitable to both GS project operators and private property owners.

6.6 A FEDERALLY COORDINATED FRAMEWORK WOULD BE OPTIMAL

The decision tree in Figure 6.2 begins with the question of whether anyone enjoys a recognized property right in subsurface pore space for GS. The answer to that question triggers a series of additional policy and legal questions. How these questions are resolved will determine whether and how geologic CO_2 sequestration will be able to move forward. Consider first the lower branch of Figure 6.2, which assumes that no vested property rights are currently recognized in subsurface pore space for CO_2 sequestration. In this case, both Congress and the states have the option of defining property rights in the pore space in some manner. In the absence of such action, sooner or later the issue will be resolved in the courts on a case-by-case basis. Judicial reconciliation could result in a patchwork of different rights from state to state. On the other hand, legislative action to fix the existence and nature of property rights would have the benefit of establishing clear and uniform principles that would yield predictable outcomes, particularly if Congress passed legislation that is applicable throughout the country. Additionally, such efforts at standardization could help to foster coordination when geologic basins underlie several different states. Alternatively, the states could enact uniform legislation—perhaps drafted under the auspices of the National Conference of Commissioners on Uniform State Laws. Uniform state laws would avoid the issue of whether a

FIGURE 6.2 *Alternative approaches to managing access and use of pore space for GS. The dashed lines represent approaches utilizing government condemnation of private property.*

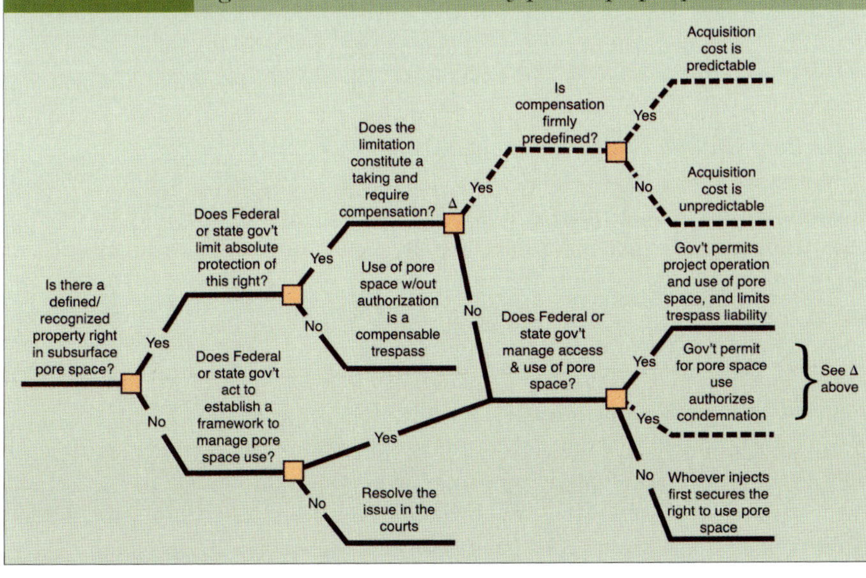

federal law would be constitutional under the Supremacy and Commerce clauses. Historically, issues of title and basic incidents of privately owned and state-owned real property have been the province of state law, while the federal government has largely passed regulatory laws. Federal law would need to be narrowly tailored to have any chance of passage and of surviving judicial review. For example, such a law might declare a public easement in deep pore space for the purposes of sequestration.

The upper branch of the decision tree in Figure 6.2 is the optimal path, especially if the law is found to have already determined the basic nature and extent of property rights to deep subsurface pore space. Both federal and state governments have intervened to regulate land use and could do the same for deep subsurface uses. Should lawmakers choose to regulate, the question becomes whether the intervention meets the constitutional standard for a taking of private property within the meaning of the Fifth Amendment to the US Constitution.[89] Assuming no taking, this approach would have the simplifying effect of leaving government with the same discretionary choices outlined above for establishing a framework for use of the deep subsurface. On the other hand, should the US Supreme Court decide (or let stand lower court rulings to the same effect) that use or regulation of pore space for GS constitutes a Fifth Amendment taking, then the government must decide (and courts must ultimately pass on the legality of) whether and how to fix the constitutionally required just compensation for such a taking. If courts determined that the degree of taking associated with using pore space at depths of a kilometer or more imposes a negligible burden on the use rights of surface property owners, the government could set compensation at a nominal level—perhaps even zero. A decision not to regulate would not, in itself, preclude sequestration projects, but developers of such projects would then be burdened with the tasks of identifying owners and negotiating contracts to secure long-term robust sequestration rights without regulatory guidance and certainty. In the West, where much of the land is federal and private holdings are often large, this might not be terribly difficult. In the East, where small, fractional ownership of real property predominates, acquiring rights for secure sequestration could entail transactions with thousands of individual landowners.

Note that several of the routes through the decision tree (Figure 6.3) result in outcomes that could make it economically difficult or impossible to develop large, commercially viable GS projects. If the federal or state governments do not intervene to manage or limit the protection of private property rights for the use of pore space for geologic CO_2 sequestration, the cost of acquiring pore-space rights could adversely affect GS project economics and greatly increase the cost of generating electricity.[90] For instance, if the use of relatively thin sandstones with low mass-to-volume storage capacities for GS result in CO_2 plumes that spread over hundreds to thousands of square-kilometers, the cost of acquiring the right to use pore space, should compensation be required, could make the project uneconomical (see Table 6.1).[91] Such costs might compare with a sequestration project's operational capital cost that, conservatively, will also be very high.[92]

FIGURE 6.3 *Approaches to managing access and use of pore space that could make development of GS projects costly and difficult.*

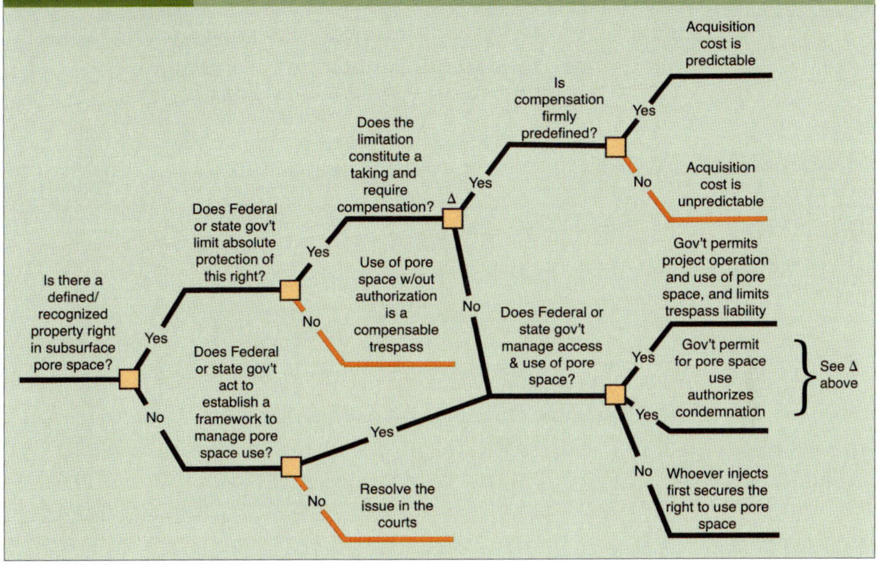

Research suggests that, if it must be paid for, the cost of acquiring the right to use pore space for GS might not be trivial and could increase the overall cost of generating electricity with CO_2 capture to an amount ranging from a fraction of a percent to up to 40%, and account for up to 20% of the total levelized cost of electricity when all GS-related costs are factored into a generation revenue requirement (see Table 6.1).[93] This could render electric generation facilities capturing CO_2 for the purpose of permanent geologic sequestration in saline aquifers uneconomic.[94] Thin formations with low porosities will likely lead to high injection costs, large CO_2 plumes, and high costs for acquiring necessary property rights, all of which will increase the levelized cost of operating a power plant with CO_2 capture and, in turn, customer electricity prices. For example, according to Gresham and McCoy, the combined annualized cost of sequestration (i.e., injection field characterization and injection facility/field capital and property rights acquisition) could alone increase the levelized cost of electricity for an integrated gasification combined cycle electric generation facility with CO_2 capture by close to $30 per MWh (Table 6.1).

If neither Congress nor the states act to establish a framework for managing access and use of deep pore space for geologic CO_2 sequestration, then resolution will be left to the courts to sort out under tort, property, and contract law. Relying on the courts to adjudicate disagreements about subsurface property rights and contractual obligations between GS project developers/operators and private property owners could significantly delay, if not permanently halt, the development of

TABLE 6.1 *CO_2 plume size estimate and present value cost (2009$) of electricity generation and GS for an IGCC facility under the imposition of a $50/t$CO_2$ price.*[1]

	Frio Sandstone (TX)	Mt. Simon Sandstone (IL)	Oriskany Sandstone (PA)	Medina Sandstone (PA)	Source
Facility Net Electric Output (MW)	530	530	530	530	CMU CEES (2010)[2]
Operational time horizon (years)	30	30	30	30	
CO_2 sequestered (metric tons CO_2/yr)	3.9×10^6	3.9×10^6	3.9×10^6	3.9×10^6	CMU CEES (2010)[2]
CO_2 emitted (metric tons CO_2/yr)	1.8×10^5	1.8×10^5	1.8×10^5	1.8×10^5	CMU CEES (2010)[2]
Levelized cost of generation ($/yr)[a]	306×10^6	306×10^6	306×10^6	306×10^6	Gresham & McCoy
Levelized cost of electricity ($/MWh)	88	88	88	88	Gresham & McCoy
Wholesale electricity price (¢/kWh)[b]	7.1	6.3	5.7	5.7	EIA (2008)[3]
CO_2 emission cost ($/yr)[c]	8.9×10^6	8.9×10^6	8.9×10^6	8.9×10^6	Gresham & McCoy
CO_2 pipeline levelized cost ($/yr)[d]	8.2×10^6	10×10^6	10×10^6	10×10^6	Gresham & McCoy
CO_2 sequestration levelized cost ($/yr)[e]	8.7×10^6	7.4×10^6	91×10^6	44×10^6	Gresham & McCoy
Pore-space lease term (years)	100	100	100	100	Gresham & McCoy
Areal extent of CO_2 plume (km^2)	440	330	4,900	2,400	Gresham & McCoy

	Frio Sandstone (TX)	Mt. Simon Sandstone (IL)	Oriskany Sandstone (PA)	Medina Sandstone (PA)	Source
Annual pore-space lease cost ($/yr)[f]	590,000 – 11×10^6	440,000 – 8×10^6	6.7×10^6 – 120×10^6	3.2×10^6 – 60×10^6	Gresham & McCoy
Generation facility revenue requirement ($/MWh)	96–99	95–98	107–123	121–152	Gresham & McCoy

Notes: a. Includes cost of capture.
b. Equals the wholesale price of electricity reported by the Energy Information Agency for the North American Reliability Corporation region corresponding to the geographic location where each of the sandstone formations are situated.
c. Based on $50 cost per metric ton of CO_2 emitted.
d. Based on 100 km pipeline length. The difference in the levelized cost of pipeline transport of CO_2 to the Frio Sandstone compared to the other sandstone formations is attributed to regional cost differences. See McCoy, S.T., and E. S. Rubin. 2008. An engineering economic model of pipeline transport of CO_2 with application to carbon capture and storage. *International Journal of Greenhouse Gas Control* 2(2): 219–229.
e. Includes annualized injection facility/field capital cost, annualized injection field characterization capital cost, and injection facility/field annual operating cost.
f. Annual lease rate range $5–100/acre/year.

Sources: 1. Gresham, R. L., McCoy, S. T., Apt, J., and Morgan, M. G. 2010. Implications of compensating property owners for geologic CO_2 sequestration. *Environ. Sci. Technol. 44*(8): 2789–2903; and Gresham, R. Lee. 2010. Chapter 4 in *Geologic CO_2 Sequestration and Subsurface Property Rights: A Legal and Economic Analysis*, doctoral thesis, Department of Engineering and Public Policy, Carnegie Mellon.
2. CMU CEES (Carnegie Mellon University Center for Energy and Environmental Studies). 2010. *Integrated Environmental Control Model Carbon Sequestration Edition Version 6.2.4*. Pittsburgh, PA: Carnegie Mellon University.
3. US Energy Information Agency (EIA). 2010. *Average Wholesale Price by NERC Region Electric Utilities and Power Marketers, 2001–2008*. http://www.eia.gov/cneaf/electricity/wholesale/wholesalet2.xls (accessed July 21, 2010).

many GS projects. This discussion of judicial barriers to GS development assumes, of course, that the appropriate property owners will be amenable to the use of the deep pore space in which they hold a vested property interest. Holdout landowners could prevent development of GS projects, especially in the eastern United States where there are innumerable small private land holdings. Absent eminent domain or unitization authority, holdouts could effectively stymie implementation of GS projects or make them very expensive, such as by bargaining for an ongoing time and volume-based storage fee. However, the cost of acquiring pore-space rights via condemnation or unitization could be economically prohibitive. Faced with highly variable and unpredictable acquisition costs, would-be GS developers might be discouraged from moving forward with a project before even attempting negotiations with the appropriate surface owners and mineral owners. Assuming the urgent need to address climate change, this is clearly an outcome that must be avoided.

Application of a Property Rule—Natural Gas Storage and Field Unitization

If pore space is determined to be protected by what is known as a "property rule," which permits invasions and use of property only with prior permission from the property owner,[95] the owner's lack of prior actual use or the lack of any reasonable, future non-speculative, investment-backed uses for the pore space would not dissuade courts from concluding that all affected surface owners (and in some cases mineral owners) have the right to exclude others from using pore space for geologic CO_2 sequestration and are entitled to compensation if the pore space is used.[96] Figure 6.4 represents this expansive view of private property rights in the subsurface in the context of CO_2 sequestration. Such would be the result if the natural gas storage model is followed. Under this approach, existing and non-speculative, investment-backed future uses of the pore space would be relevant only in determining the amount of just compensation due.[97] This approach may be a natural extension of the Montana, North Dakota, and Wyoming legislation, which declare that surface owners hold title to subsurface pore space.[98] Thus, this approach would most likely require compensation to surface owners for any CO_2 sequestration, either by bilateral agreement or through the exercise of eminent domain.

Alternatively, compulsory unitization legislation similar to that used to facilitate the enhanced recovery of oil and gas could be adapted for GS. Unitization for GS would essentially be a contractual instrument backed by statutory authority that

FIGURE 6.4 *Property rule approach to managing access and use of pore space for geologic CO_2 sequestration.*

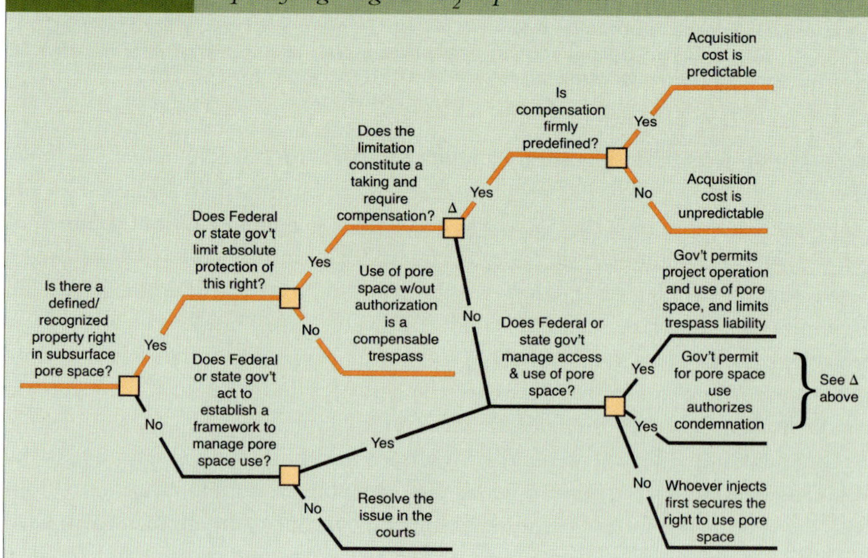

requires landowners to consent to the use of their pore space for CO_2 sequestration if the majority of the surface acres (based on percentage) within a regulatory-approved project boundary had voluntarily committed to be developed as a single sequestration unit. Note that compulsory unitization, in the context of a GS project, is not fundamentally different from eminent domain. The difference lies in the fact that a statutory percentage of surface owners would have to agree voluntarily to the establishment of a GS project before non-consenting owners could be forced to transfer sequestration rights for just compensation through a regulatory proceeding that is, practically speaking, similar to a court proceeding in eminent domain.

The Interstate Oil and Gas Compact Commission (IOGCC) developed a model statute for GS that contemplates the acquisition of property rights in the same manner as for achieving compulsory unitization for EOR or, alternatively, through eminent domain as is done for natural gas storage projects. They write, "The Model General Rules and Regulations propose the required acquisition of these storage rights and contemplate the use of state natural gas storage eminent domain powers or oil and gas unitization processes to gain control of the entire storage reservoir."[99] This is an extension of a known regulatory program. The IOGCC Model Statute has the advantage of working through the well-established mechanisms of state oil and gas agencies that are familiar with drilling and reservoir regulations. However, the IOGCC Model Statute also has shortcomings. One is that it does not authorize CO_2 sequestration to be conducted in formations containing economically recoverable amounts of oil and gas, or other valuable resources.[100] Under the IOGCC's model, a GS developer could not acquire (by either unitization or eminent domain) the right to use a formation that had once produced minerals without first establishing, by agreement or otherwise, that the minerals are exhausted. Over the large areas that could be affected by a CO_2 sequestration operation, this may be difficult to prove. What is more, the question of just how depleted mineral-bearing strata must be before the pore space can be used for CO_2 sequestration may be addressed differently from state to state and over time. Some states view the issue of mineral exhaustion very favorably to the interests of the mineral estate owner, taking into account potential new production methods and technologies.[101] This means that if a GS project was developed and it was later discovered that CO_2 was being injected into, or had migrated into, strata containing recoverable resources, the only option might be to discontinue sequestration operations. Thus, project developers may well be motivated to identify formations, such as saline aquifers, that have never yielded valuable minerals and have little prospect of doing so.

There is a good chance that the need to acquire property rights in the subsurface from the required threshold of landowners—perhaps hundreds or thousands—to develop a GS site will make the IOGCC Model Statute's unitization model administratively unwieldy and economically unattractive. This model requires a GS project developer to identify and negotiate in good faith with all property owners "having property interests affected by the storage facility."[102] Consequently, the unitization model would likely result in higher costs associated with acquiring the pore-space right necessary for geologic CO_2 sequestration than would be realized under the more simple eminent domain approach that limits required compensation to only those instances where the injection and migration of CO_2 materially impairs current or non-speculative, investment-backed future uses of the subsurface. An expansive view of subsurface property rights modeled after the unitization or, for that matter, under the natural-gas-storage eminent domain model that appears to entitle all affected landowners to compensation, regardless of any actual harm, could greatly discourage the development of GS due to the potentially overwhelming cost of having to compensate all property owners overlying the CO_2 sequestration reservoir.[103] With CO_2 sequestration, this could be true even if the value of just compensation is clearly defined and constrained, as in *Loretto*,[104] because the CO_2 plume could migrate over hundreds to thousands of square-kilometers.[105] Of course, the nature and extent of the money expended and the infrastructure needed will depend on how widely GS is deployed, where it is deployed, and how integral a techno-strategy GS becomes in America's approach to limiting GHG emissions.

Federal or State Ownership of Pore Space

Just as *Causby*[106] confirmed that federal legislation cut off property interests in the higher airspace, legislation authorizing the use of the deep subsurface for CO_2 sequestration could similarly truncate property interests in the deep subsurface, except in connection with those uses that are currently in existence or subject to non-speculative, investment-backed expectations, just as West Virginia's CCS working group recommended. Numerous courts have held that a surface owner's interest in the subsurface is "limited" at best, relying on *Causby* and other cases limiting the surface owner's right to control the airspace.[107] Arguably, even if states expressly provide by statute that a surface owner has a property right in the pore space, as Wyoming, North Dakota, and Montana have done, such a state-created property interest may be limited by the judicial application of *Causby* to subsurface rights that places "objective" limits on rights to the subsurface.[108] In other words, the argument would be that just as Wyoming could not vest in surface owners the right to the airspace far above their property as a result of the objective, background principles expressed in *Causby*, Wyoming cannot vest in surface owners the right to the deep subsurface as a result of courts' application of *Causby* to the subsurface.[109]

This argument is consistent with a 2008 article by John Sprankling in the *UCLA Law Review* entitled "Owning the center of the Earth"[110] in which he argues that private property rights to land should not extend more than 300 meters below the surface of the earth and that the subsurface beneath that threshold should belong to the federal government.[111] The article does not focus on geologic CO_2 sequestration specifically, but instead discusses the issue of subsurface ownership in connection with today's technological ability to develop various energy and climate change technologies, including GS, that must make use of the subsurface in ways not contemplated in the past.[112] Sprankling contends that based on case law involving subsurface water, oil and gas development, and hazardous waste injection, among others, American law has never determined whether a landowner's rights extend more than three kilometers below the surface, and that even case law within three kilometers of the surface is largely inconsistent.[113] He concludes that property owners should have some rights below the surface to accommodate foundations, trees, and other normal surface facilities, but those rights should not extend more than 300 meters below the surface.[114]

Following Sprankling's argument, Congress or the states could enact legislation declaring a public easement exists in subsurface at a specified depth below the surface of the earth.[115] Such an "easement" would not need to transfer ownership of subsurface minerals to the federal government, and the easement could even be expressly subordinate to mineral rights but require accommodation[116] by mineral owners to minimize conflicts. The legislation could also establish a system for compensating property owners with existing uses of the subsurface below that depth if they are harmed by CO_2 injection, and truncate the establishment of future private property rights and expectations going forward. This regime would need to formalize and standardize procedures for authorizing access and use of pore space for CO_2 sequestration. Such a framework would obviate many of the property rights conflicts that might arise when CO_2 sequestration projects involve the use of pore space in more than one state. Figure 6.5 shows the path federal or state governments could take in establishing ownership and control of deep pore space for the purpose of CO_2 sequestration.

FIGURE 6.5 *Federal or state ownership approach to managing access and use of pore space for geologic CO_2 sequestration.*

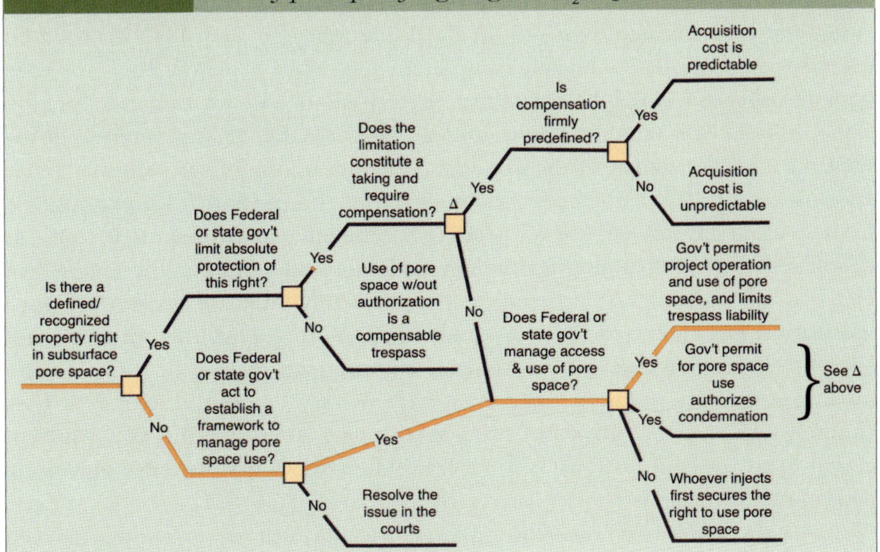

While such an approach would facilitate the development of CO_2 sequestration by simplifying the process of accessing the right to use pore space for GS and constraining the cost of acquiring subsurface property rights, it would almost certainly invite takings challenges. More importantly, it is doubtful that it would be politically feasible for the Congress or state legislatures to implement such a legal framework, which would certainly depend on the political climate and attitudes of the courts over the coming decades.

A Liability Rule Limiting the Protection to Planned and Existing Uses

While vesting deep subsurface property in the federal or state governments has appeal as a means to facilitate new technologies like GS, such an approach fails to recognize the realities of how the subsurface has historically been used and is used today. In many regions of the country, subsurface property rights below 300 meters include coal production, oil and natural gas exploration, production, and storage, freshwater production and storage, fluid waste and wastewater injection, and compressed air energy storage. Congress has chosen to implicitly recognize those property rights under some circumstances, such as through the eminent domain provisions of the Natural Gas Act,[117] and courts have recognized those rights by allowing for claims of trespass and nuisance in cases of actual interference or harm.[118] Courts also have created mechanisms to compute just compensation when subsurface areas are needed for a public use such as natural gas storage.[119] Thus, the country's history of the use of the subsurface is in fact different from its use of the airspace.

Even though common law and statutory authority prevent the vesting of property in airspace other than when it is used in connection with the surface, the same is far less true with regard to the deep subsurface. To date, there has been no federal declaration of a "public highway" in the subsurface in the way there has been with airspace, and, at depths of less than several kilometers, any future declaration along those lines would come into conflict with vested economic interests in the subsurface in many areas of the country.[120] Economic use of the subsurface may end at a certain depth (for instance, any deeper than is necessary for existing and future natural gas storage, waste injection, and oil and gas development), but this depth is subject to changing technology and economics. And to the extent that CO_2 sequestration will be at depths that are currently subject to existing or non-speculative, investment-backed uses (and it appears that it will be), there do not appear to be any background principles of common law that would prevent states from vesting those property rights in surface owners or mineral owners if they choose to do so, or prevent courts from recognizing contrary rights as a matter of common law or constitutional law—at least where the courts have not previously determined ownership.[121]

An approach based on existing and non-speculative, investment-backed uses would likely result in the protection of subsurface property rights in some circumstances but not in others, based on whether the geology is suitable for CO_2 sequestration, as well as whether sequestration might compete with oil and gas development, natural gas storage, and the like. Protecting subsurface property rights based on existing uses and non-speculative, investment-backed expectations would provide a middle ground approach to property rights that makes geologic CO_2 sequestration somewhat more expensive to implement, but would recognize, value, and compensate for competing economic uses that would be impaired by GS, and consequently lessen opposition to such projects. Moreover, this approach is firmly grounded in common law precedent. In the context of enhanced recovery of oil and gas, fluid waste injection, and freshwater storage and recovery, the courts have refused any absolute protection of property rights in the deep subsurface, but have retained limited protection of property rights that would allow property owners to recover monetary compensation for damage to property caused by actual and substantial harm or interference. Allowing recovery only for actual damage to property is different from finding that a landowner possesses the type of property right in the subsurface that empowers the landowner to prevent others from injecting fluids into the pore space underlying the landowner's property (i.e., a property rule); it is this type of absolute protection of subsurface property rights that the courts seem to have clearly rejected in the context of enhanced hydrocarbon recovery, underground fluid waste injection, and freshwater storage.

One application of this standard, shown in Figure 6.6, is in essence a "first-in-time, first-in-right" approach to the use of pore space for CO_2 sequestration,

Figure 6.6 *"First-in-time, first-in-right" approach to managing access and use of pore space for geologic CO_2 sequestration.*

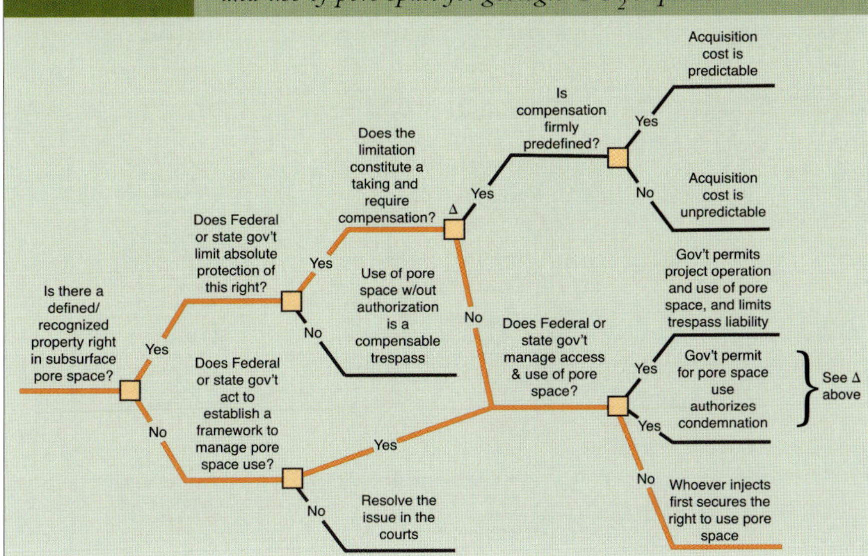

where neither the government nor its agencies would oversee and manage the right to access and use subsurface pore space for CO_2 sequestration. Under this approach, a GS operator would possess the privilege to inject CO_2 into subsurface pore space with the knowledge that it will migrate through the targeted geologic strata provided it is in compliance with regulations covering injection operations. Of course, the uncompensated use of pore space would only be permissible if it does not interfere with a verified existing or non-speculative, investment-backed use of the subsurface that has been asserted by a property owner—this is what is known as a "liability rule."[122] The use of pore space for underground disposal of fluid wastes is governed by a liability rule.

A second, and our recommended, application of this standard would be for the federal government to codify a formal process for managing the access and use of pore space for geologic CO_2 sequestration, whereby the project developer acquires a permit to use the pore space for GS from the appropriate UIC permitting agency (see Figure 6.7). While delegation of implementation to the states, as under the UIC program, would be possible, a strictly state-based framework would have difficulty addressing problems related to sequestration under federal lands, sequestration in geologic reservoirs that cross state lines, and other interstate issues related to the transport and injection of CO_2 into the subsurface on the scale required to meaningfully address climate change.

FIGURE 6.7 *Recommended approach to managing access and use of pore space for geologic CO_2 sequestration.*

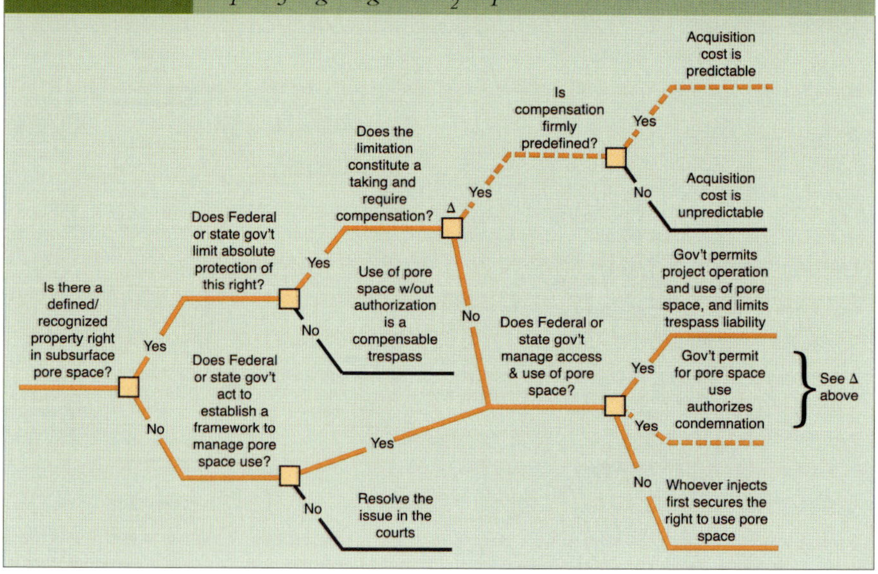

6.7 AUTHORITY TO PERMIT GEOLOGIC CO_2 SEQUESTRATION ON FEDERAL LANDS

The Department of Energy's Carbon Sequestration Atlas estimated that around 5.5% of the US onshore geologic CO_2 sequestration capacity lies beneath leasable federal lands.[123] Sequestering CO_2 on federally owned lands has two advantages over sequestering on private land: fee simple ownership of surface, mineral and pore space rights, and large contiguous tracts of land. However, a fair proportion of the roughly 250 million surface acres and 700 million acres of subsurface mineral estates managed by the United States Bureau of Land Management (BLM) is intermixed with private land. Therefore, any permitting framework for GS on federal lands must address two confounding property-related issues: 1) the situation where ownership of the surface and mineral estates is split between the federal government and private landowners (i.e., "split estates"), and 2) sequestering CO_2 on marginally sized tracts of federal land held in fee simple by the government that are bordered by privately owned lands. Use of federal land will also require compliance with all applicable federal, state, and local laws and regulations that protect environmental and human health, including the National Environmental Policy Act[124] (NEPA), the SDWA and the EPA's UIC regulations, the National Historic Preservation Act[125] (NHPA), and the Endangered Species Act[126] (ESA). NEPA will require extensive analysis and documentation of the potential environmental impacts of GS projects proposed on federal lands.[127] Areas of Critical Environmental Concern (ACECs), such as wilderness areas, national parks and monuments, will not be available for GS development.[128]

Current federal statutes and regulations do not directly address injection of CO_2 into the subsurface on federal lands for permanent geologic sequestration. The Federal Land Policy and Management Act[129] (FLPMA) and the Mineral Leasing Act[130] (MLA)—which give the Bureau of Land Management (BLM) the authority to issue rights-of-way (ROWs) for pipelines and other infrastructure, the licensing of oil and gas exploration for production, the injection of CO_2 for enhanced recovery of oil resources, and the injection and underground storage of natural gas (all on federal land)—could easily be modified to specifically authorize the use of federal lands for geologic CO_2 sequestration.[131] Both Congress and the Department of the Interior (DOI) have examined the FLPMA and the MLA to determine to what extent the DOI and the BLM can, and should, exercise authority over GS on federal lands, as well as what changes to public land laws are necessary.

The Federal Land Policy and Management Act

The FLPMA provides ample legal authority to approve geologic CO_2 sequestration projects on BLM-administered federal land. Title V of the FLPMA authorizes the BLM to issue rights-of-way over and under federal land for a variety of systems, including: 1) systems for the transportation of storage liquids and gases (other than natural gas or synthetic gaseous fuels); 2) systems for the generation, transmission, and distribution of electric energy; and 3) any other systems or facilities that are in the public interest and require rights-of-way.[132] Title V rights-of-way, which can be easements, are issued for a term that gives the holder the certainty and security needed to obtain adequate financing, even if this means that the term is "very long or even perpetual."[133] Congress intended that BLM's authority under Title V "be all inclusive and provide [the BLM] the requisite authority to grant any right-of-way for any purpose that is in the public interest."[134] Title V was "intended to include rights-of-way which serve future needs arising out of existing and future technology advances. This clause is broad enough to cover rights-of-way for ... any other systems which are not yet in general use."[135] Accordingly, there is little question that the BLM's authority under Title V is broad enough to cover geologic CO_2 sequestration.

Even if the BLM's authority under Title V is determined not to cover geologic CO_2 sequestration, Section 302(b) of the FLPMA grants the BLM the authority to permit CO_2 sequestration on federal lands. This provision authorizes the BLM to issue easements, permits, and leases for industrial and commercial uses that cannot be authorized under other laws.[136] In fact, in 2009 the DOI submitted a report to Congress that recognized Section 302(b) could be used for geologic CO_2 sequestration.[137] The report stated that the BLM has the statutory authority to allow federal land under its jurisdiction to be used for geologic CO_2 sequestration under either Title V or Section 302(b) of the FLPMA. Even so, the FLPMA should be amended in a manner so as to expressly convey to the BLM the authority to issue rights-of-ways and licenses for construction and use of pipelines and other systems required for the transportation, distribution, and permanent geologic sequestration of CO_2 on federal lands.

The Mineral Leasing Act

The MLA authorizes the Secretary of the Interior to approve the subsurface storage of natural gas "in [federal] lands leased or subject to lease" to "avoid waste and promote conservation of natural resources," regardless of whether the gas is produced on federal lands.[138] In *Exxon Corp. v. Lujan*,[139] the United States Court of Appeals for the 10th Circuit affirmed a BLM decision to issue a right-of-way for a CO_2 pipeline under the MLA rather than the FLPMA.[140] While the case is not entirely on point because it dealt with the issue of CO_2 pipelines and not natural gas storage, it provides some meaningful precedent. The MLA right-of-way provisions govern oil and natural gas pipelines, while the FLPMA provisions govern pipelines other than oil and natural gas pipelines. The court reviewed Exxon's appeal of the BLM permit issuance, specifically about whether a CO_2 pipeline could be characterized as a natural gas pipeline under the MLA. The *Exxon* court upheld an earlier ruling by the Federal District Court of Wyoming, which found that the MLA never defined the word "gas" and that the plain meaning of "natural gas" was ambiguous with regard to whether it encompassed CO_2. Because of the statutory ambiguity, the district court looked to the legislative history of the MLA.

After analyzing the Congressional debate of the precursor to the MLA, the court found that if one were to categorize gases in the broadest possible manner at the time the MLA was enacted, "they would fall into two categories: natural gas—that is, gases that occur naturally—or artificial gas; namely, gases manufactured in the laboratory."[141] The district court further found that if Congress had wanted to define natural gas restrictively in the MLA, Congress knew of the term "hydrocarbon" and could have defined "natural gas" to mean "gaseous hydrocarbons," excluded smaller components of the natural gas such as CO_2, or simply used the term "hydrocarbon."[142] The court also referred to a legal opinion from the Department of the Interior's Office of the Solicitor, which argued that the MLA refers only to "gas" or "natural gas" without any qualifying adjectives, and that a non-restrictive reading of the terms would be supported under the oil and gas leasing provision of the MLA.[143] The *Exxon* court then held that the term "natural gas" had a "technical meaning thus precluding reliance on its ordinary definition."[144] Any use of the world "natural" was meant to distinguish the gas from that which was "artificially produced."[145] Extending the *Exxon* court's analysis of the MLA's right-of-way provisions to subsurface storage provisions, it would appear that CO_2 falls within the MLA's definition of "gas," and that geologic CO_2 sequestration would be governed by the MLA's provisions on the "subsurface storage of natural gas."

6.8 RECOMMENDATIONS ON ACCESS TO PORE SPACE

Permanent geologic sequestration of CO_2 and access to pore space for GS should be permitted under a modified version of the EPA's UIC program. Implementing a federal framework for permitting the access to and use of pore space for GS will require federal legislation specifically aimed at addressing and promoting geologic CO_2 sequestration, as well as amendments to the Safe Drinking Water Act and a modification and expansion of the Class VI UIC permitting rules for GS promulgated by the EPA.[146] One distinct advantage of grafting a framework for managing the access to and use of pore space for CO_2 sequestration onto a modified version of the UIC program is that such an approach could capitalize on the experience of UIC regulators (either at the EPA or in designated state environmental and natural resource departments or oil and gas commissions),[147] their relationship with state environmental and natural resource agencies and geologic surveys, and the knowledge about GS UIC regulators amassed while preparing the rules for regulating the injection of CO_2 for permanent geologic sequestration. Another advantage to this approach is that, with modified authority that allows for such balancing, UIC regulators would be in a good position to balance the need for permanent geologic sequestration of CO_2 in response to climate change with the need to protect underground sources of drinking water supplies so as to ensure the overall safety of human health and the environment.

Currently, the regulatory context for geologic sequestration is shaped by a strong history of state and local control of land use issues, property rights issues, and significant variation among states in both geology and applicable law. State environmental and natural resource agencies, oil and gas commissions, mining agencies, and geologic surveys, all of which have invaluable local knowledge, should play a significant role in permitting geologic sequestration projects as either the EPA-designated UIC regulator or consultants to the UIC permitting process for CO_2 sequestration. Collectively, state regulatory agencies are well positioned to promote the development of GS, protect underground sources of drinking water, and prevent the degradation of economically recoverable hydrocarbon resources and hard mineral resources. However, while private property rights have historically been the province of state law, federal law will need to address the acquisition of pore space rights on federal lands, split estates, and interstate lands.

Statutory Declaration that Geologic CO_2 Sequestration is a Public Use

Federal legislation must declare that permanently sequestering CO_2 in deep geologic formations to combat the detrimental effects of climate change is a public use undertaken in the national interest. Furthermore, federal legislation should create the presumption that the regulatory grant of the right to access and use pore space for geologic CO_2 sequestration does not amount to a compensable taking because the issuance of a permit to use pore space 1) does not effect a confiscation of property, and 2) is not the first step in a regulatory taking since pore space owners will unlikely suffer either an actual loss or an interference with any investment-backed expectation. However, property owners should be afforded the opportunity to rebut this presumption by presenting evidence in an administrative permitting proceeding which demonstrates that CO_2 sequestration will result in a material impairment to a current or non-speculative, investment-backed future use of the subsurface, and that the property owner will suffer a consequent economic loss requiring just compensation.

Federal Legislation Authorizing UIC Permitting Agencies to Permit Access and Use of Pore Space for Geologic CO_2 Sequestration

Federal legislation should establish procedures under which UIC regulators are authorized to issue pore space permits granting GS project operators, who have received a ClassVI CO_2 injection permit under the UIC program, the exclusive right to access and use pore space for geologic CO_2 sequestration within the subsurface project boundary[148] specified by the injection permit. This is not to suggest project developers should be prohibited from acquiring pore space rights through private negotiations. On the contrary, if a project developer acquired consent to use pore space from all property owners overlying the subsurface project boundary defined in the CO_2 injection permit, a pore space permit would not be necessary.

Requirements for States Seeking Primary Permitting Authority for Geologic CO_2 Sequestration

As with the existing UIC program, states and tribes should be able to apply for primary enforcement responsibility to implement the UIC program for CO_2 sequestration within their borders. However, states and tribes seeking primary enforcement responsibility under the UIC program for geologic sequestration of CO_2 should be required to develop a framework and enact statutory authority for acquiring the necessary ownership interests, easements, or licenses to occupy subsurface pore space. State UIC programs for GS should be further required to take into account the effects that permanent geologic sequestration of CO_2 in the permitting state will have in any other state. In the same vein, federal UIC regulations must ensure that any state with a reasonable prospect of being affected by the issuance of a CO_2 injection permit under another state's UIC program has the right to intervene and participate in proceedings conducted by the permitting state. If a GS project will require the use of pore space located in more than one state (i.e., an interstate project) with primary UIC enforcement authority, the relevant states should be required to enter into a cooperative agreement with respect to permitting and regulating the project. If the states fail to reach an agreement, the EPA should assume primary enforcement authority of the interstate project.

Competitive Application Procedure for Class VI CO_2 Injection Permit

When a geologic CO_2 sequestration project developer applies for a Class VI CO_2 injection permit, a sufficient period of time should be afforded to other GS project developers and operators who wish to contend for the issuance of a CO_2 injection permit and, if applicable, argue that a pore space permit will impair their own ability to develop and operate an alternative and potentially competing geologic CO_2 sequestration project. Such parties should have the right to intervene and file a competing application. An intervener should be entitled to equal consideration with the original applicant if they file a competing application within a specified time period after public notice of the original application is filed.

Application Procedure for Pore Space Permits

If a CO_2 injection permit applicant seeks to develop a geologic CO_2 sequestration project in a state (or states) that has (or have) not codified statutory authority for accessing and using pore space for CO_2 sequestration, and the applicant has not acquired through voluntary contract the ownership interests, easements, or licenses necessary to occupy pore space, they should submit a CO_2 injection permit application. Then, a sufficient period of time for participation in the permit application proceeding should be afforded to all interested parties and owners of preexisting interests that may be materially impaired by the granting of the pore space permit. If an interested party or preexisting interest-holder fails to intervene within a period of time specified after public notice of the original application was filed, such party should be deemed to have waived any and all rights and property interests that may become impaired by the geologic CO_2 sequestration project should a pore space permit be issued.

If it is demonstrated that a preexisting interest would be materially impaired by the issuance of a pore space permit, the geologic CO_2 sequestration project should be permitted only upon 1) a modification of the project that avoids the impairment; 2) a contractual agreement between the owner of the preexisting interest and the project applicant; or 3) a finding by the UIC regulator that condemnation of the preexisting interest through the exercise of eminent domain, with appropriate compensation, is necessary to the proper operation of the CO_2 sequestration project under application.

Subsurface Trespass Liability Related to Geologic CO_2 Sequestration

Project operators injecting CO_2 pursuant to a valid CO_2 injection permit, and, if necessary, a pore space permit, should be protected from trespass liability. Specifically, legislation or the courts should decree that the use of pore space should not give rise to an actionable trespass claim unless actual and substantial harm is caused to established protectable property interests. A federal standard for subsurface trespass liability would obviate the need to make a federal declaration or modify state common law and statutory determinations regarding pore space ownership. Instead, the role of determining ownership of pore space should remain with the states. A pore space permit should not protect a GS operator if injection or migration of CO_2 materially impairs preexisting interests that are identified during the UIC permit proceeding, or property interests outside the project boundary specified by the CO_2 injection permit.

Creation of Eminent Domain Authority for Geologic CO_2 Sequestration

Federal legislation authorizing the use of eminent domain for geologic CO_2 sequestration is necessary as a backstop in situations where the right to use pore space cannot be acquired through voluntary contractual negotiations. This need could arise for projects which, having been deemed by a UIC regulator to be in the public interest, are being unreasonably delayed or thwarted by owners of protectable property interests that are identified during the UIC permitting process, or when absolute control of the sequestration field is required to protect the integrity of the field. If use of eminent domain is warranted, the holder of a CO_2 injection permit should be granted the right, via the issuance of a pore space permit, to appropriate any subsurface strata required for the project to move forward.

Licensing Procedure for Geologic CO_2 Sequestration on Federal Lands

The BLM and EPA should jointly license geologic CO_2 sequestration projects on federal lands and split estates following the UIC permitting procedures laid out in the discussions above in Section 6.5. Additionally, the BLM should issue perpetual easements under Title V of the FLPMA for geologic CO_2 sequestration exploration and development projects on federal lands through a two-phase licensing procedure: one license would be issued to study and characterize the proposed CO_2 sequestration site, the other to develop the site and conduct CO_2 injection. Application of this framework would enable a GS project to be permitted on federal and private lands simultaneously.

Phase I: Project Area License

Geologic CO_2 sequestration project developers should have to apply for a project area license, which would grant the applicant the exclusive right to conduct environmental and geological testing and monitoring within the proposed subsurface project boundary. Priority of application should be given to the first complete application the BLM receives. The project area license term should be for a period of five years, during which time the pore space within the proposed project boundary should not be available to other potential applicants. During this first phase, the holder of a project area license must formulate a Plan of Development (POD) for the proposed geologic CO_2 sequestration project to be submitted during the second phase of the application process. A plan of development would be a detailed design of the geologic sequestration project and CO_2 transportation, distribution, and injection facilities. The plan of development helps the BLM assess the public safety and environmental effects of the proposed CO_2 sequestration project and its facilities.

Phase II: Project Development License

To qualify to apply for a project development license for CO_2 sequestration on federal land, the holder of a project area license must submit a plan of development to the BLM no later than five years from the time the project area license was issued. A project developer must simultaneously apply for a CO_2 injection permit for geologic sequestration at the time it files for a project development license with the BLM. A project development license should be awarded based on the merits of the applicant's plan of development as well as be contingent on the applicant securing a CO_2 injection permit for geologic CO_2 sequestration. A project development license, in combination with a CO_2 injection permit and, if necessary, a pore space permit, would grant the applicant the right to construct, operate, and maintain underground wells and necessary surface and subsurface facilities for the permanent geologic sequestration of CO_2 within the project boundary specified in the UIC permit. Project development license proceedings and CO_2 injection permit proceedings for projects that will be carried out on split estates will be subject to the requirements outlined in Chapter 4 if the project applicant has not acquired the ownership interests, easements, or licenses necessary to occupy pore space on private lands through voluntary contract or state statutory authority.

Environmental Review of Federal Actions Authorizing Geologic CO_2 Sequestration on Federal Lands and Split Estates

The BLM should be the lead agency for the purpose of federal environmental review pursuant to the National Environmental Policy Act and, in consultation with all relevant agencies, should prepare a single environmental review document for each license application phase to be used as the basis for decisions under federal law related to each project phase. The environmental review for the project area license should only assess the likely environmental effects of the testing facilities to be constructed and operated under the terms of the license. The environmental effects of permanent geologic sequestration of CO_2 should not be considered for environmental review of the project area license. Instead, the environmental review for the project development license should assess the likely environmental effects of geologic CO_2 sequestration and the facilities and wells constructed for that purpose.

Summary

Although the EPA has the authority under the UIC program to permit and regulate geologic sequestration of CO_2, the agency currently has no authority to consider the subsurface property issues attendant to permanent CO_2 sequestration. There is a possibility that CO_2 sequestration operations could interfere with other current or foreseeable uses of geologic pore space. A legal framework for GS is needed that balances the interests of property owners with the public benefit of CO_2 sequestration and reduces the possibility of interference with other commercial uses of the subsurface. We recommend that the federal government codify a formal process for managing the access and use of pore space for CO_2 sequestration in the form of an amendment to the UIC program under the SDWA. Under the system we propose, a project developer would acquire a permit to use the pore space for

GS from the appropriate UIC permitting agency only after other parties (that may have active or non-speculative, investment-backed future uses of that space) have had an opportunity to indicate their interests, and potential conflicts have been resolved. In certain cases, the grant of a permit to use pore space should also convey the right to invoke eminent domain authority when it is necessary to ensure that the integrity of the sequestration field is not compromised. Additionally, legislatures or the courts should establish a subsurface trespass standard such that the use of pore space for GS by a valid pore space permit holder is not compensable unless the owner of the pore space or other affected property owners suffer actual and substantial damages. Such a framework should increase the potential for either avoiding disputes of subsurface use or resolving them at the outset in a stable and predictable manner, as well as constrain the cost of acquiring pore space rights.

7

LIABILITY AND THE MANAGEMENT OF LONG-TERM STEWARDSHIP

Two of the most important questions that must be answered if CCS is to become a large-scale, commercially viable means of reducing US CO_2 emissions are: 1) What will be the liability of GS project operators (and other entities) for personal injury, property damage, trespass, and nuisance claims that arise over the course of the lifetime of a geologic sequestration project? 2) What is the appropriate institutional framework for managing GS sites after closure?

7.1 LIABILITY ACROSS A PROJECT'S LIFE CYCLE

Liability rules are important not just because they determine whether, when, and to what extent parties will be compensated for any injuries or losses they may bear if a GS project goes awry, but also because investors' and public perceptions about likely future liabilities strongly affect their reactions to projects currently under consideration. Liability issues that arise in connection with capture facilities will be similar to those that arise for any large industrial facility such as a power plant or refinery. Liability issues involving CO_2 pipelines will be similar to those that arise with natural gas pipelines. For this reason, we focus here on liability issues related to GS.

In addition to making it clear how losses will be compensated, liability rules are also important because without clear and workable liability rules that permit project operators to assess their risk exposure, project developers may not be willing to pursue commercial-scale GS projects that can meaningfully reduce CO_2 emissions from coal-fired power plants and other major sources of CO_2.[1] Thus a liability regime for GS will need to satisfy at least three public policy objectives: 1) reinforce incentives for prudent siting, operation, and closure of GS projects; 2) assure availability of compensation for economic losses to the public in case of property damage or personal injury during the period of operation and after closure; and 3) provide clear rules for and an efficient system for adjudicating compensation claims and assessing financial responsibility for payment of these claims.

GS projects that inject and sequester CO_2 in deep geologic formations pose some unique problems for determination of liability and allocation of risk that vary according to the phase of operation of the project (see Figure 1.7). Each phase carries a different set of risks that require a different approach to management and treatment of liability.

The nature and extent of liabilities posed during the pre-operational, operational, and immediate post-operational phases will be familiar to GS project operators and insurance companies.[2] For example, the risks and potential liabilities are expected to be similar to those experienced during underground injection of fluid wastes and during various oil and gas production operations, particularly enhanced oil recovery.[3] These similar and familiar activities suggest that risks, and any associated liabilities, will be highest during the project's operational phase.

At the same time, because the risks posed during the pre-operational, operational, and immediate post-operational phases are likely known and quantifiable, it appears that commercial insurance will be available at reasonable costs.[4] For this reason, it does not appear that changes to the current liability regime will be required for the GS projects in these phases (except for "first mover" projects, which are discussed later in this chapter). The same situation will apply to a site that is closed but still being managed by the site operators, prior to its being transferred to another entity for long-term stewardship. Hence, in the balance of this chapter, we focus on the issue of how liability should be addressed during long-term stewardship. Of course, the answer depends on how long-term stewardship is managed, so before turning to the details of choosing a liability regime, we first discuss the strategy we propose for managing long-term stewardship.

7.2 DESIGNING A STRATEGY TO MANAGE LONG-TERM STEWARDSHIP

There are basically four alternatives for managing GS sites once they have closed, become stable, and are ready to enter long-term stewardship: 1) continued management by the GS project operator; 2) management by some other private party; 3) management by state government; or 4) management by the federal government. The odds are low that any specific well-designed site will subsequently become a problem, require extensive remediation, and incur costs to damaged or injured parties. However, at least today, it appears that no commercial insurance companies are willing to write policies to cover this contingency. If separate funds are set aside (e.g., in an escrow account) to cover the full range of contingencies that might occur at each individual injections site, or the set of sites in a single state, the amount of money that would need to be tied up would be much larger than if the risk could be spread nationwide. If a state took on this responsibility, if a major problem developed, and if the fund was not adequate to cover remediation and compensation, Congress would likely be called upon to bail them out, with ultimate adverse consequences for citizens overall.

Additional challenges could arise with regard to jurisdiction if the sequestration site crossed state lines. Issues would likely arise in determining the share of risks among states involved in selection of interstate geologic formations.

There is also the problem that "long term" may mean thousands of years before sequestered CO_2 becomes mineralized. Private firms cannot be counted on to exist that long, nor is it possible to ensure that any legal obligations they have will be honored many years in the future. For both these reasons, we conclude that long-term stewardship should be handled by a federal entity. Federal responsibility carries the added benefit that it levels the playing field for investors in different states, and helps to provide consistent oversight.

A site should not be handed off, or accepted, for long-term stewardship before it is stable and poses minimal risk. It seems prudent to not have the same entity that is responsible for regulating the operation of a site (i.e., the EPA or a state under the UIC program) also be responsible for the decision to accept a site into long-term stewardship. Accordingly, we recommend the creation of a Federal Geologic Sequestration Board (FGSB). This would be an independent board, for administrative convenience housed in the Department of Energy, with the technical capability to determine whether a GS site is sufficiently stable to be accepted into long-term stewardship. The FGSB would oversee routine surveillance of all sites that had been accepted into long-term stewardship; would be able to call upon appropriate parties to perform remedial activities at a site should that become necessary; and would set risk-based fees that active GS projects would pay into a fund controlled by the FGSB to cover the administrative costs of long-term stewardship, ongoing monitoring, and any possible necessary site-remediation, and the compensation of damages. To assure rapid resolution of any problems at operating sites, the FGSB would also be authorized to draw upon the fund for emergency remediation of major accident events, with subsequent recourse back to the operators and commercial insurers.

7.3 TYPES OF LIABILITY THAT MAY ARISE DURING LONG-TERM STEWARDSHIP

Potential liabilities during long-term stewardship fall into three broad categories: tort liability, climate liability, and regulatory liability. (Another category, contract liability, is traditional in considerations of liability. However, outside its GHG and climate context, it can be dealt with by existing legal mechanisms.)

Tort liability includes the obligation to pay compensatory or punitive damages if the injected CO_2 causes harm or injury to others during the long-term stewardship phase. Examples could include damages if injected CO_2 harms drinking water or impedes the exploitation of mineral rights. The potential magnitude of tort liability during the long-term stewardship phase is highly uncertain, which of course is why, today, commercial insurance companies will not write policies.

Climate liability is the obligation to pay compensation for GHG emissions if injected CO_2 leaks back to the atmosphere. Climate liability is a form of contract liability in that CO_2-generating entities will avoid statutory—or regulatory—imposed GHG emission penalties by permanently sequestering CO_2, and leakage would be a breach of that contract. As discussed in Chapter 8, the form of compensation for climate liability will depend on the specifics of the statutory GHG reduction program. For example, under a cap and trade system, it might entail submitting emission allowances. Chapter 8 discusses the forms of compensation for climate liability within the other potential GHG reduction programs. The potential magnitude of climate liability is very hard to estimate due to uncertainties about the future price of carbon, lack of knowledge and experience with long-term leakage rates, and lack of clarity about the timing and effectiveness of remediation to stop leaking emissions.

Regulatory liability is the obligation to pay for post-operational activities required by regulation. Tort and climate liability are both examples of the common understanding of "liability"—that is, a legally enforceable obligation to compensate a person or entity that has suffered a discrete harm that they were otherwise entitled to avoid. Although technically also a "liability," regulatory liability is fundamentally different from tort and climate liability in one respect: it is a legal obligation that must be fulfilled regardless of negligent behavior or harm. In this sense, it is like being "liable" for payment of taxes, even if one has not broken any laws or societal standards. The lower limits of regulatory liability will be established by rules that set monitoring and reporting requirements during long-term stewardship and, therefore, are easily quantifiable. The upper limits are more difficult to estimate because they will be associated with potential remediation costs, which may be site-specific. The Federal Geologic Sequestration Board that we have proposed will need to do an assessment to first set the fee that operators will deposit per tonne of injected CO_2. On a periodic basis, this fee may be adjusted as experience accumulates. Once more is learned, it may even be possible to define site-specific fees that are risk or performance based.

Regulatory liability costs for monitoring and maintenance will be incurred for every site during the long-term stewardship phase. For most sites, regulatory liability costs for remediation and mitigation will not be incurred, but if they become necessary they could be very high.

An effective liability regime for the long-term stewardship phase must consider who should bear each type of liability—tort, climate, and regulatory—and how those types of liability can best be managed.

Regimes for addressing liability can be thought of as lying along a spectrum, as shown in Figure 7.1 and elaborated in Box 7.1 at the end of the chapter. Importantly, under the strategy we propose, the choice to make a federal entity (the FGSB) responsible for long-term stewardship does not mean that the federal

FIGURE 7.1 *Alternative approaches to dealing with liability can be thought of as falling along a spectrum of alternatives, as shown in this figure. Significantly, as discussed in the text, the fact that the federal government assumes the responsibility for long-term stewardship need not mean that it assumes all liability. Rather, a mixed strategy seems best, depending on the nature of the issues that arise.*

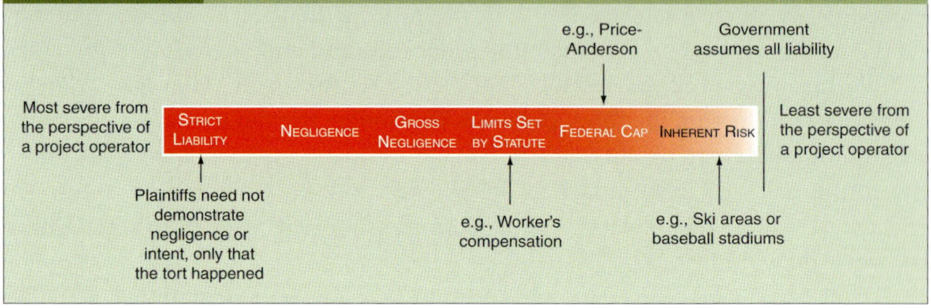

treasury will be responsible for covering the costs associated with tort, climate, or regulatory liability.

7.4 A HYBRID APPROACH TO LIABILITY DURING LONG-TERM STEWARDSHIP

In order to assure that the operating costs of long-term stewardship do not fall on the US Treasury, the FGSB would review the proposal for each GS project and assess a fee per tonne of sequestered CO_2. This review will be done independently of, and subsequent to, the EPA UIC process of issuing a site license. In Chapter 4 we outlined the benefits of an adaptive approach to regulation. While the FGSB should establish a base fee for all projects, over time, as more is learned about how to assess proposed reservoirs and their subsequent performance, it may be possible to make at least a portion of this fee risk-based. By collecting fees from sites operating all across the nation, the FGSB will create a system of pooled liability.

Because any funding mechanism must remain solvent over an extended period (centuries), determining the appropriate level of the fee is important. In addition to performing probabilistic risk assessment, factors that will need to be considered in setting a stewardship fee during the first decades of operation should include: 1) the rate of injection; 2) the date when the first GS projects will likely enter the long-term stewardship phase; 3) the requisite value of the fund at the time when the first projects go into the long-term stewardship phase; 4) the rate at which the fund value must increase as additional projects enter the long-term stewardship phase; and 5) the investment yield for trust funds.

While income from this fund should be sufficient to cover the costs of FGSB operations and ongoing periodic monitoring of sites that have been accepted for long-term stewardship, it should also be sufficient to cover the cost of any low probability, but potentially expensive remedial activity that might be required. Note that, if there were no federal system of the sort we propose and a problem developed with a site that involved very large expenses under a long-term stewardship program being operated by a private entity or by a state, there is a high probability that the federal government would have to assume some significant responsibility (see the discussion of Florida hurricane insurance in Box 7.1). A nationally pooled, operator-financed fund is better equipped to handle any contingencies that might arise, with far lower risk to the Federal Treasury, while also moving more rapidly to address the problem.

In applying to move a site into long-term stewardship, a GS operator would supply the FGSB with full details on site characterization and performance. If, after the site was accepted into long-term stewardship, a problem were to develop and it was found that some of those details had been mischaracterized or excluded, the operator would not escape liability. Notwithstanding the general release from liability that occurs once stewardship begins, the project operator would remain liable to the Board for any compensation paid or remediation cost incurred by the FGSB as a result of the project operator's willful failure to comply with any regulatory requirement or any false or misleading statements made in connection with any stewardship certificate application.[5] The FGSB would move rapidly to begin remediation while also seeking recovery from the former operator, or its successor(s).

In addition, project operators would remain subject to claims under express contracts, including any contractual liability to CO_2 producers for atmospheric leakage under a cap and trade program. Monitoring and remediation costs, as well as any compensation claims, are paid out of the fund. If the amount in the fund were insufficient to pay costs and claims, the FGSB could increase assessments to current and future operating projects or request an interest-bearing advance from the Treasury. The Treasury would establish terms and conditions to ensure repayment of the advance.

There is some history of federal fund management to address tort liability.[6] One example is the Trans-Alaska Pipeline Liability Fund (TAPL Fund), now part of the funding available under the Oil Pollution Act of 1990 (OPA) (see discussion in Box 7.1). This system provides an important analog for GS for at least two reasons. First, the OPA reconciles existing regulatory standards and incorporates approaches to liability and risk management depending on the location of the damage. Second, it creates a significant fund for quick payout of claims in case of harm but allows claimants to seek damages in excess of the fund's maximum from liable parties under state or federal tort law. These features make the liability structure for claims associated with the Trans-Alaska Pipeline (and now oil spills in general) particularly relevant for GS.

The FGSB would set compensation standards and adjudicate all compensation claims (see discussion of capped and limited liability in Box 7.1). Compensation standards and the FGSB's administrative procedure should be designed to provide reasonable and prompt compensation for economic losses in connection with a closed project (on the basis of "claims made"). The FGSB would have jurisdiction over these claims subject to judicial review in federal courts. The FGSB's compensation standards could depart from common-law tort rules as it determined appropriate. Any remediation at stewardship sites that occurred would be pursuant to FGSB rules, but either the FGSB or the EPA might order emergency action.

This proposal strikes a balance between providing incentives for appropriate behavior, ensuring compensation for any damages arising from GS facilities, and shaping a risk profile that will encourage investment in GS facilities.

It is very important to assure that the long-term stewardship fund is not subject to discretionary appropriation and will not have to compete with other discretionary spending. Hence there must be a firewall between this fund and other treasury funds or it must be independently administered to prevent appropriations issues. This feature is especially important because of the long delay between the time the fund is created, the time industry participants pay into it, and the time the money will be required for compensation and maintenance. Several options can accomplish this protection: 1) classifying the fees as an "off-setting collection" which would enable it to be credited against appropriations; 2) authorizing direct expenditures from the fund without appropriations; 3) locating the fund "off budget," which would allow it to be exempt from appropriations and excluded from fund limitations in federal budget rules.

The FGSB would be authorized to retain private contractors to perform routine or emergency operations. It would also be authorized to delegate stewardship responsibility to state agencies and to enter into contracts to reimburse them for any costs of monitoring or remediation that they incur. However, the Board would remain responsible for compensation.

7.5 FIRST-MOVER PROJECTS

First-mover projects are GS demonstration projects or other GS projects that begin injection, or go to financial closing, before the program for commercial projects is in place. A first-mover GS project would be subject to otherwise applicable state and federal liability rules during the operational phase and would be expected to rely on private insurance. After project closure, however, the Department of Energy would be authorized, although not required, to enter into long-term agreements to indemnify the project sponsor for all or part of potential liabilities of the long-term stewardship phase. These commitments could be entered into in advance of appropriations, up to specified maximum levels. The Department of Energy, in consultation with the EPA, could impose conditions designed to protect health, safety, and the environment, as well as secure the financial interest of the government. Once the federal stewardship program for commercial projects becomes operational, first-mover projects would be transitioned into, and be covered by, those programs.

7.6 RECOMMENDATIONS ON HOW TO ADDRESS LIABILITY AND LONG-TERM STEWARDSHIP

We make the following recommendations with respect to how liability should be treated across the life cycle of a GS project and for the institutional arrangements to manage long-term stewardship.

Operating commercial GS projects should remain subject to liability rules under otherwise applicable state and federal law and should rely on the private insurance market, or mutual insurance, for risk management.

A federal program operated by a Federal Geologic Sequestration Board (FGSB) should be created to oversee the long-term stewardship of adequately closed injection projects. The FGSB should be an "independent agency," but housed within an existing federal agency, such as the Department of Energy, for purposes of administration. It should administer, and be financed by, a revolving fund that is based upon risk-based assessments on GS projects during their operating life.

Once a GS project is completed and has been demonstrated to be stable and well behaved, the FGSB will review an application for moving it to long-term stewardship. The FGSB will do this by assessing the project in terms of the standards the FGSB has established and confirming that it does not present unreasonable health, safety, or environmental risks. Having done this, the site should be transferred to the federal long-term stewardship program along with all liability and responsibility for compensation. However, the site operator or its successor(s) would not be relieved of liability that results from their willful failure to comply with any regulatory requirement or any false or misleading statements made in connection with any stewardship certificate application.

Necessary remediation or compensation payments during the stewardship phase should be the responsibility of the FGSB, and should be disbursed from the revolving fund. The FGSB could also make the fund available for emergency remediation of sites not yet covered by the long-term stewardship program (with the prior requirement that the FGSB will recover costs of remediation from the project operator or other parties).

Depending on the timing of general federal CCS legislation, it may be necessary to establish a stop-gap federal indemnity program for the stewardship phase of "first-mover" projects. Those projects should ultimately be covered by the stewardship program outlined in this book.

Establishing the legal and institutional framework for long-term stewardship is essential to ensuring that, many decades hence, a publicly governed and financially responsible entity will be available to monitor closed GS sites, undertake any necessary remediation, and provide the public with assurance that compensation will be paid for damage to persons or property.

Box 7.1 *Explanation and Elaboration of the Several Strategies for Addressing Liability that Lie Along the Spectrum in Figure 7.1.*

Absent federal or state statutory changes, the state tort law system, administered by state and federal courts, provides the framework for determining liability. This system governs compensation for civil wrongs, including personal injury, property damage, trespass, and nuisance. State tort systems vary from state to state because they are amalgams of common-law rules and statutory provisions established over several centuries.

Receiving compensation under tort law often requires an injured party to prove three elements: 1) harm occurred; 2) the defendant breached an applicable standard of care; and 3) the breach caused the harm. Because of these burdens of proof, the tort system works best to ensure compensation for injuries resulting from discrete and easily traceable causes. In other circumstances, however, the tort system can be expensive, inefficient, and the cause of significant uncertainty, particularly when the discrete cause of the harm is unclear, such as in areas of high scientific complexity.

Thus, the potential for tort liability can affect business behavior when the threat of tort liability deters businesses from engaging in socially beneficial activities that are also potentially dangerous. Conversely, for less risk-averse firms, a system of pure tort liability may not sufficiently encourage responsible behavior because the cost of litigation might be less than implementing precautionary measures, but high enough to deter injured parties from seeking compensation.

Within tort liability itself, there is a spectrum of potential liabilities, ranging from severe to virtually non-existent liability.

Strict liability imposes an absolute duty to avoid harm and does not require an injured party to demonstrate that the injury resulted from another party's unreasonable action or intent to cause harm. Strict liability is typically reserved for ultra-hazardous and inherently dangerous activities such as the storage of hazardous waste. Products liability cases are also often litigated under a strict liability standard. Strict liability, or some version thereof, offers significant protection to injured parties and encourages prudent operation. However, it can also lead to expensive and lengthy litigation, as has occurred in the hazardous waste context, where litigation has slowed the rate of waste cleanup.

Negligence and gross negligence are the next entries as we move from left to right along the illustrated spectrum of strategies for addressing liability. Under a negligence rule, an injured party must show that the defendant owed the injured party a duty of care, that they breached that duty, that the breach was a proximate cause of the injury, and that the plaintiff suffered actual harm. No showing of intent, poor motive, or aberrant activity is required in order to assess liability under an ordinary negligence scheme. In contrast, under a "gross negligence" standard, the burden of proof for injured parties is much greater because it requires the same showing as ordinary negligence, *plus* proof that the defendant consciously or recklessly disregarded the duty to the injured party.

At both the "negligence" and the "gross negligence" points on the spectrum, defendants have several affirmative defenses that they can raise to avoid or to limit liability. Those defenses include contributory negligence, comparative negligence, proximate cause, and assumption of risk. Contributory negligence bars a plaintiff from recovery if the defendant demonstrates that the plaintiff contributed to the harm suffered through the plaintiff's own negligence. Comparative negligence reduces the amount of damages a plaintiff can recover based on the degree to which the plaintiff's own negligence contributed to the cause of the injury. Asserting a lack of proximate cause means the alleged cause of an injury is too remote from the harm suffered. Assumption of risk bars a plaintiff from recovery if the defendant can demonstrate that the plaintiff voluntarily and knowingly assumed the risk inherent in the dangerous activity in which s/he was participating at the time of the injury. For example, assumption of risk, when applied to activities such as skiing, relieves ski slope owners and operators from liability for injuries that occur through standard use of their facilities.

Private liability includes third-party insurance, bonding, and internal company funds that assure the availability of resources. Private liability is best suited for well-understood risks over finite timeframes.

Insurance is a contract between two entities in which one entity agrees to reimburse the other entity for losses suffered in exchange for a consistently paid premium. Insurance serves to pool and allocate risk through classifying the risk and pricing it, using policy exclusions and deductibles, and through creating "surrogate regulation," where private parties perform or require "governmental-like" actions such as inspection, risk assessment, and risk management in order to act as the impetus for better management.[7] Insurance is functionally a form of privatized risk pooling, but pooling can also occur by statute (discussed below). In the insurance context, liability is pooled by the private insurance company.

Insurance has long been used to manage environmental risk, both operational and catastrophic.[8] The Resource Conservation and Recovery Act (RCRA) and the Comprehensive Environmental Response, Compensation, and Liability Act (CERCLA) use pollution liability insurance as a tool to control environmental damages. As a general matter, commercial insurance is only available if the following five conditions are satisfied. First, a sufficient number of similar and uncorrelated events must exist to allow for risk pooling. Second, projects must have clearly calculable losses. Third, the time period for potential losses must be well-established. Fourth, losses must be frequent enough to calculate premiums. Fifth, the insured party must not have an incentive to cause losses.

Bonding is another strategy to address liability. A bond is a written promise to pay money if specified circumstances occur or after a specified time elapses.[9] A relevant example of bonding is provided by the EPA's UIC program. All injection wells regulated under the UIC program, and most state-regulated oil and gas production wells, require bonding to help ensure proper site closure. In the UIC program, an operator must submit a well closure and abandonment plan that identifies steps for closing the well (plugs, cement, cost) and any subsequent post-closure monitoring activity.[10] While a performance bond is required to ensure proper plugging and abandonment, in most cases no long-term monitoring is legally required, and the bond is released upon well closure. For UIC wells, the bond would be released after the operator has satisfied plugging and abandonment procedures established by the regulator. Bond amounts are established by the states and differ significantly across jurisdictions.

Statutorily modified liability regimes have been implemented through statutes enacted by state legislatures and Congress to achieve a desired policy end. These statutes alter the tort and private liability regimes discussed above, creating statutorily modified liability regimes.[11]

Statutory schemes can cap the amount of financial exposure and liability a particular entity faces. A liability cap is a statutory enactment that limits the amount of liability claims the entity must pay to those affected by some potentially disastrous event, such as an oil spill or nuclear power plant accident. Liability caps are generally appropriate when the government wants to encourage investment in nascent technologies that present a low probability of extremely high-cost accidents or events. Often, these caps explicitly couple limits on damages for liability without fault. A cap on liability typically takes into account the average or likely cost of addressing harms associated with the particular activity. Supplemental mechanisms for compensating injuries could help address injuries or financial losses in excess of the cap and assuage any public concern about a cap on liability.

At the federal level, the Price-Anderson Act,[12] discussed further in the following "pooled" and "tiered" liability sections, limits the damages an injured party may collect from operators of nuclear power plants as a result of a nuclear incident. Specifically, the Act bars punitive damages and imposes a cap on total damages. Plant operators must obtain individual plant liability insurance and must participate in an industry-wide retrospective rating plan to provide a second tier of liability coverage. The Act immunizes nuclear power plant operators from public liability for amounts that exceed the required private liability insurance.[13] Should damages exceed this level, Congress may provide additional compensation to protect the public.[14]

The Trans-Alaska Pipeline Authorization Act (TAPAA)[15] created a liability cap for injuries resulting from the transportation of oil along and from the Trans-Alaska Pipeline.[16] Strict liability applies up to the level of the cap, but injured parties can recover damages in excess of the cap only if the responsible parties acted negligently.[17] In another example involving oil, the Oil Pollution Act of 1990 (OPA) limits liability for oil spills. The OPA caps liability at $75 million, but the limit does not apply to incidents resulting from an operator's "gross negligence or willful misconduct,"[18] and it does not displace tort liability under other state or federal laws. The recent Deepwater Horizon oil spill raised public concern about the OPA's liability cap, putting pressure on Congress to remove the limit altogether.[19] Indeed,

BP, who leased the offshore rig, stated that it would waive the $75 million statutory cap on liability for economic damages. BP did not, however, admit to any gross negligence or misconduct related to the spill. At the same time, BP established a $20 billion fund to pay claims resulting from the spill, which an independent administrator oversees.[20] In the wake of the Deepwater Horizon incident, any regime that caps liability for projects with potentially harmful environmental consequences will likely face close public scrutiny.

Pooled liability includes private pooling—insurance—and statutorily created risk pooling. Risk pooling can occur through statutes that modify tort liability and attempt to mitigate the extreme nature of strict liability by spreading the risk across an entire pool of owners and operators.[21] This alternative to the tort process relies on a pooled fund to pay claims, often with a set cap on damages available from the fund. The fund is usually financed by contributions from the entities likely to need the fund. Each entity's required contribution can be set based on any variety of factors.

Workers' compensation laws incorporate a liability cap, discussed above, and a pooled liability regime to displace tort liability and compensate employees injured in the workplace.[22] Based on a statutory schedule, the program compensates injured workers directly but precludes liability adjudications in court. Employer-funded insurance or pooled employer contributions fund the compensation program. These compensation programs pool liability in order to average out the variability of individual damage awards. This compromise allows for limited damages for liability without fault, enabling injured workers to recover damages while limiting the uncertainty that businesses face.

Under the Price-Anderson Act, damage claims that exceed the amount of private insurance available to a nuclear power plant are paid from an industry-funded pool, into which each plant operator contributes up to approximately $120 million if an accident occurs—that is, the fund is not paid into unless an accident occurs. In the event of an accident, operators are required to pay roughly $17.5 million annually until the claim is met or the maximum is reached. Currently, the nation's 103 operating nuclear power plants are liable in the aggregate for contributions of approximately $11.6 billion to the fund.

Congress also created pooled liability to support the Vaccine Injury Compensation Program (VICP), which compensates victims of vaccine-related injuries.[23] This program provides a no-fault alternative to the traditional tort system. An excise tax of seventy-five cents on all vaccines, deposited in the Vaccine Injury Compensation Trust Fund (VICTF), finances the program.[24] In the Vaccine Program, parties who suffer an injury listed on an enumerated table may receive compensation through a trial process in which the injured party's burden to demonstrate that the vaccine caused the injury is lower than it would be under a traditional negligence theory of liability. In an innovative design, regardless of the claim's success, reasonable attorney's fees are also paid to those bringing claims.[25]

The Oil Spill Liability Trust Fund (OSLTF) provides another model of a government fund to pay liability claims. Established in the OPA, the fund provides up to one billion dollars per accident. A fee of five cents per barrel of domestic or imported oil financed the fund from its creation in 1990 until a sunset provision took effect in 1994. The fee was reinstated in the Energy Policy Act of 2005. The sunset in 1994 of the OSLTF's major funding mechanism proved extremely problematic, and in the twelve years between 1994 and 2006 it was significantly underfunded. Currently, the funds for OSLTF come from investment interest accrued, cost recovery from parties responsible for oil spills, and fines and civil penalties collected under the OPA, the Federal Water Pollution Control Act, the Deepwater Port Act, and the TAPAA.[26] The fund is administered by the US Coast Guard's National Pollution Funds Center.

The reason that Congress has relied upon pooled liability so often is very simple: it offers a middle ground, mitigating financial burdens on potentially liable parties while still providing support to victims.[27] This compromise can prevent liability from becoming overly burdensome. Proponents insist that this limitation is necessary to protect potentially liable parties from crippling financial penalties, which encourages investment and innovation in risky fields. However, detractors argue that pooled liability can weaken the incentive to strive for good performance by breaking the link between failure and a high risk of penalty, thus creating a "moral hazard."

Tiered liability modifies tort liability and increases protection for injured parties while limiting the financial burden on an individual company or industry. In addition to the liability cap, discussed above, the Price-Anderson Act applies tiered liability to all existing U.S. nuclear power plants.[28] Congress enacted the Price-Anderson Act to encourage development of civilian nuclear energy by limiting tort liability for nuclear accidents and guaranteeing that citizens could be compensated for damages to people and property. The Act creates tiers of financial responsibility, consisting of private insurance, an industry pooled fund, and a cap on total liability. 10 CFR §140.83 limits liability from potential "extraordinary nuclear occurrences." 10 CFR §140.11 requires every nuclear reactor over ten megawatts to have $375 million in insurance. Additional claims are paid from the industry-funded pool discussed above. Like pooled liability funds, tiered liability is controversial. Proponents argue that the Price-Anderson Act has been crucial for nuclear industry development and has obligated the nuclear plant operators and the industry to maintain a higher level of liability insurance coverage than might otherwise occur.[29] Critics say the Act serves as a public subsidy to the nuclear industry and limits the ability of affected parties to recover adequate compensation.[30]

Indemnification models involve one party compensating another party for loss or otherwise providing security against loss or injury.[31] Indemnity typically involves a contractual agreement or legal obligation to cover such loss. This model is typically used for liability to third parties. When used, the injured party has to pursue the claim against the indemnified person in order to be compensated. Therefore, indemnification does not necessarily ensure that the injured party will be compensated.

Indemnification is a component of private insurance (discussed above). Insurance companies indemnify policyholders against loss covered by the policy. The federal government can use this model to limit liability by agreeing to indemnify a manufacturer or service provider against loss that the entity incurs. This model is often combined with a cap on liability or some form of pooled fund that offsets all or part of the cost to the public entity providing indemnification.

Government assumption of liability is the final entry illustrated on the spectrum in Figure 7.1. For example, in Florida, many insurance companies concluded that the liabilities that they assumed in coastal communities were greater than they could manage, given rate limits that the state imposed to encourage economic development. Accordingly, Florida established its own insurance program, Citizens Property Insurance Corporation (CPIC), to give consumers an affordable option. This approach is not without risks, however. In 2007, Citizens went from being an insurer of last resort to being the largest insurer against wind losses in Florida. If Florida is hit by a catastrophic hurricane in the next few years, it is very likely that CPIC will not have sufficient funds to cover its losses. If this happens, the state will have the legal obligation to assess policyholders, issue bonds to cover the cost, tax its citizens in some fashion, or seek assistance from the federal government in order to back up the CPIC.

8

GREENHOUSE GAS ACCOUNTING FOR CCS

In the world of finance, the term "accounting" refers to a system for recording and summarizing business transactions and then analyzing, verifying, and reporting the results. In this chapter, we review the systems that will be needed to analyze, verify, and report the technical performance of CCS and link that performance to a climate or energy policy. The purpose of greenhouse gas (GHG) accounting is threefold: 1) evaluate the overall effectiveness of CCS as a GHG emissions reduction technology; 2) accurately measure its value as a GHG emission reduction technology, allowing it to be compared with other options; and 3) protect the integrity of the climate or energy policy.

The basic point of CCS technology is to allow the continued combustion of fossil fuels without emissions of CO_2 to the atmosphere. If and when a GHG emissions reduction program is created that has rules that allow emissions reductions from CCS, then GHG accounting protocols will be vital to document both the amount of CO_2 that is sequestered and the CO_2 emissions avoided by using CCS (see Box 8.1).

Box 8.1 *What is the Difference Between the Amount of CO_2 Sequestered and the Amount of CO_2 Avoided?*

The terms "emissions avoided" and "emissions reductions" require a point of comparison or baseline. That necessitates an answer to the question: what would have happened in the absence of the CCS project?

At first glance, the answer seems simple: the amount of emissions captured and successfully sequestered equals the amount that would otherwise have gone into the atmosphere, and thus, is the amount for which credit should be given. However, this answer ignores the emissions that result from the extra energy needed to power those processes used to capture the CO_2, and it also ignores the fact that electricity might have been procured from some source other than CCS; for example, wind or natural gas. Hence, the baseline for comparison may be extremely difficult to estimate due to its dependence

The Difference Between the Amount of CO_2 Sequestered and the Amount of CO_2 Avoided

		Emissions Avoided by Substituting CCS (kg CO_2/MWh)[b]	
Potential Baselines for Comparison	**Emissions Rate (kg CO_2/MWh)[a]**	**SCPC w/CCS**	**NGCC w/CCS**
Super critical pulverized coal w/CCS	107[c]	0	64
Natural gas combined cycle w/CCS	43[c]	–64	0
Super critical pulverized coal (SCPC)	811[d]	704	768
Natural gas combined cycle (NGCC)	367[d]	260	324
Low-carbon (e.g., wind, solar)	0	–107	–43
US average	603[e]	496	560
Washington State average	150[e]	43	107
West Virginia average	875[e]	767	832

Notes: a. This is the average emissions rate for either the generation technology or the noted region; these estimates include only the emissions from combustion of fossil fuels (or lack thereof). Including emissions over the entire fuel cycle would increase these emissions rates. See Jaramillo, P., W. M. Griffin, and H. S. Matthews. 2007. Comparative life cycle air emissions of coal, domestic natural gas, LNG, and SNG for electricity generation. *Environmental Sci. and Technol.* 41: 6290–6296.

b. Positive values of avoided emissions mean that substituting CCS decreases GHG emission relative to the baseline. Negative values of avoided emissions mean that substituting CCS increases GHG emissions relative to the baseline.

c. See, for example, Kartha, S., M. Lazarus, et al. 2004. Baseline recommendations for greenhouse gas mitigation projects in the electric power sector. *Energy Policy* 32(4): 545–566; and Steenhof, P. 2009. Estimating greenhouse gas emission reduction potential: An assessment of models and methods for the power sector. *Environmental Modeling and Assessment* 14(1): 17–28.

d. Rubin, E. S., C. Chen, et al. 2007. Cost and performance of fossil fuel power plants with CO_2 capture and storage. *Energy Policy* 35(9): 4444–4454.

e. US Environmental Protection Agency. eGRID2007 Version 1.1.

on the structure of the electric industry (e.g., competitive markets versus a regulated monopoly model), the time period of interest, and the mix of generation types in the relevant region being examined.[1] The emissions "avoided" using CCS could be calculated on a project basis, as compared to the generation of the same amount of electricity from a coal-fired plant without CCS, a natural gas plant, or a wind farm. Alternatively the emissions could be calculated at the level of the regional electric grid, compared to the average emission intensity of all generation or the emission intensity of the marginal unit of electricity displaced by electricity generated at a plant using CCS. The table above illustrates how "avoided emissions" can depend on different electricity sources and different regions of the country.

On the other hand, measuring CO_2 emissions in isolation requires no such comparison—from an accounting perspective, CO_2 is either emitted or it is not, and by definition CO_2 that is sequestered is not emitted. For this reason, GHG accounting systems that can operate without requiring the definition of a baseline, such as cap and trade or performance-standard based systems, will be more accurate and less complex than systems that require the estimation of "avoided emissions."

A range of technical and policy questions must be considered when establishing GHG accounting protocols for CCS. Technical questions at capture sites focus on the efficiency of the capture process, or the "energy penalty," which will influence both the cost of capture and the GHG emission rate of electricity produced using CCS. Technical questions at GS sites largely focus on the risk of leakage and capabilities of monitoring techniques. Policy questions are concerned with fairly attributing the economic value of CCS, evaluating its effectiveness in reducing GHG emissions, and comparing it with other GHG mitigation methods. The policy options are constrained by technological feasibility, cost, measurement accuracy and precision, funding stability, and intergenerational equity. In this chapter, we focus on GHG accounting systems for CCS associated with electricity production; however, most of the same principles will also apply to geologic sequestration of CO_2 captured at other types of industrial facilities. Of course, it is also important to consider potential emissions that could result if leakage were to occur at a GS site. Indeed, this is one of the most complex aspects of GHG accounting for CCS. The following sections frame the issues in GHG accounting for CCS, review relevant current rules, assess technical considerations for monitoring and measurement, investigate options to compensate for potential surface leakage at GS sites, discuss the implications of various climate or energy policies for CCS accounting protocols, and then present our recommendations for GHG accounting protocols for CCS.

8.1 FRAMING THE ISSUES IN ACCOUNTING

GHG accounting systems will need to fit within the likely structure(s) of the CCS industry. CCS will typically involve three components: 1) the capture facility, such as a power plant that captures CO_2; 2) transport infrastructure, which moves CO_2 from the site of capture to the sequestration site; and 3) the GS project, where CO_2 is injected into the subsurface (see Figure 1.7 and the discussion in Chapter 2). Each of these components can be owned and operated by separate commercial entities, and in the case of the transport infrastructure and the GS project, may handle CO_2 from multiple capture facilities. While vertically integrated CCS operations may develop, GHG accounting procedures must be able to function effectively and provide the correct incentives when each component of a CCS "project" is separately owned.

A successful CCS GHG accounting system must include three tiers of monitoring, measurement, and reporting. The first tier is the facility level, where each operation associated with CCS should be required to report GHG emissions. The second tier is the inter-facility level, with CO_2 inventory tracking to verify the transfer of CO_2 from capture facility to pipeline operator to GS site. The third tier is the GHG reduction program level, which consolidates facility and inter-facility reports, as well as data from other GHG emitters, in order to enforce the program rules.

Facility-level GHG Emissions Accounting

Each entity associated with the CCS system, including CO_2 capture facilities, CO_2 pipelines, and GS sites, should monitor, measure, and report their GHG emissions. For GS sites, GHG emissions could come from equipment leakage (or intentional venting) and from surface leakage[2] (CO_2 that escapes the injection zone, travels all the way to the surface and enters the atmosphere). Monitoring and measurement of equipment leakage is relatively straightforward, but this is not true for surface leakage. While experience with monitoring techniques to detect and quantify surface leakage of CO_2 is growing, significant uncertainties remain about capabilities and costs. These technical questions and related policy issues are discussed in Section 8.3.

Monitoring that is solely for the purpose of GHG emissions reporting will be a relatively small part of the overall monitoring program at a GS site. The primary role of monitoring will be to assure that the project does not pose serious risks to health or the environment and is meeting its compliance obligations under the EPA UIC program. No surface leakage is expected at well-managed geologic sequestration projects undertaken at appropriate sites. However, because there will always be uncertainty about deep geologic formations, the possibility remains that unidentified geologic or man-made features (such as abandoned wells that have not been adequately sealed) could allow slow leakage of CO_2 to the atmosphere at some sites. Regulations should be designed to encourage operators to avoid surface leakage, and if it does occur, ensure that it is measured, reported, and "paid for" by the operator. Policy options to structure financial responsibility for surface leakage are discussed in Section 8.4.

CO_2 Inventory Tracking

The second level of GHG accounting should be designed to verify the transfer of CO_2 between the capture facility and pipeline operator and from the pipeline to GS site. Each business that handles the CO_2 should be responsible for monitoring and measuring the amount of CO_2 they receive from and deliver to others. Much like package delivery, there will be commercial incentives to document these transfers. But, in addition to providing proof of delivery for the commercial entities involved in the transaction, the CO_2 inventory tracking system should also require each business that handles CO_2 to report inter-facility transfers to the government agency that oversees the GHG reduction program. These reports should then be reconciled to verify that carbon dioxide captured at a power plant or industrial facility actually reaches an appropriate geologic sequestration site and is injected for permanent sequestration.

GHG Reduction Program Level Accounting

Any climate or energy program will have at least three objectives: 1) ensure adequate energy is available; 2) reduce GHG emissions in a cost-effective way; and 3) encourage development of sustainable, secure energy sources. Large and small program design decisions will struggle to manage tradeoffs between these objectives and others. The GHG reduction program will set rules that limit GHG emissions and create a value for low-carbon energy technologies. The government agency in charge of the GHG reduction program will consolidate and reconcile facility GHG emission reports and CO_2 inventory tracking reports in order to assure that all parties are meeting their compliance obligations and protect the integrity of the GHG emission reduction program. Program level data will also be used to assess the performance of the CCS system as a whole.

A future GHG reduction program could take many forms including a carbon tax, a cap and trade program, a portfolio standard that requires a certain mix of low-carbon energy sources, a performance standard or a technology standard. The form of the GHG reduction program will structure the financial costs and benefits associated with CCS, and thus influence the structure of the CCS industry and the competitiveness of CCS as a low carbon energy technology. Section 8.5 reviews the implications that different climate or energy programs would have on protocols for GHG accounting for CCS.

The early CCS industry may have strong ties with enhanced oil recovery (EOR), but any future GHG program will need to take care in its treatment of CO_2 that is injected for the purposes of EOR. Verifying the permanence of sequestration will be vital to the integrity of the climate or energy program, and EOR operations do not currently include this type of monitoring. GHG accounting protocols for CCS will need to differentiate between CO_2 injection operations that include monitoring and measurement to verify permanent sequestration and those that do not.

8.2 CURRENT RULES RELEVANT TO GHG ACCOUNTING FOR CCS

As of early 2011, the US has no federal program capping GHG emissions. Limited GHG reductions may be required under new EPA rules that make some new large facilities (or facilities that make major modifications or construct large additions) subject to permitting requirements for greenhouse gas emissions under the Clean Air Act.[3] Under these rules, GHG emissions would be reduced using Best Available Control Technology (BACT) analysis. BACT analysis for GHG permitting will set a site-specific technology or emission rate standard by choosing the most effective available method for reducing GHG emissions, taking into account technical, energy, environmental, and economic factors.[4] The EPA's white paper on available and emerging technologies for reducing GHG emissions from coal-fired electric generating units states that the agency considers CCS to be an available technology, but then suggests that concerns about costs and integration of the capture, transportation, and sequestration elements make it unlikely that CCS would be selected as BACT.[5]

Despite the present lack of an effective federal climate program, elements of a GHG accounting system for CCS are emerging. In 2010, the US EPA Office of Air promulgated rules for the Mandatory Reporting of Greenhouse Gases Program. The EPA made clear that "This rule does not require control of greenhouse gases, rather it requires only monitoring and reporting of greenhouse gases."[6] The EPA's mandatory greenhouse gas emissions reporting rule (MRR) requires reporting from upstream suppliers of fossil fuels and industrial gases, as well as downstream emitters of GHGs.[7] The rule covers many of the activities associated with CCS and EOR, including:

- supply of petroleum products;
- supply of coal-based-liquid fuels;
- supply of natural gas;
- supply of carbon dioxide;
- electricity generation.

Amendments to this rule, finalized in December 2010, require emissions reporting from petroleum and natural gas systems (Subpart W),[8] and from carbon dioxide injection and geologic sequestration (Subpart RR).[9]

The MRR program requires some, but not all, of the types of businesses that could be part of the CCS system to monitor, measure, and report their GHG emissions. The EPA estimates that facility-level GHG emissions reporting will cover 90% of emissions from electricity generation, 85% of total oil and gas industry emissions, and 60% of emissions from ethanol production.[10] Any facility that captures CO_2 and exports it for GS or EOR must report the mass of CO_2 captured and exported. At present CO_2 pipeline operators have no reporting requirements, but GS facilities and EOR operations do, as detailed below. The MRR also establishes some, but not all, components of a CO_2 inventory tracking system, by requiring CO_2 capture facilities, GS sites, and EOR operators to report the mass of CO_2 captured, extracted, imported, exported, or injected.

The MRR requirements for GS site GHG emissions reporting create thorough facility- and inter-facility-level accounting protocols during site operations. All GS facilities must report the source of CO_2 they receive, the mass of CO_2 transferred onsite, and the mass of CO_2 injected. They must also monitor and report emissions in three categories: vented CO_2, equipment leakage, and surface leakage. Each GS site must develop a site-specific monitoring reporting and verification (MRV) plan, and revise that plan based on site performance as necessary. The plan must include:

- leakage risk assessment—identification and characterization of all potential leakage pathways;
- monitoring strategy—a site-specific plan that may include a combination of subsurface, vadose zone, surface water, and/or atmospheric monitoring, and modeling. Continuous atmospheric monitoring is not required. Leakage must be quantified if CO_2 is detected at the surface;
- pre-injection environmental baselines—a site-specific strategy to establish pre-injection CO_2 levels;[11] and
- site-specific mass balance equations to calculate the amount of CO_2 sequestered.

MRR reporting requirements for EOR facilities do not create accounting systems adequate to verify that CO_2 is permanently sequestered. EOR facilities must simply report the source of CO_2, the mass of CO_2 transferred onsite, and the mass of CO_2 injected. EOR facilities are specifically required to report emissions from EOR injection pump blowdown and CO_2 dissolved in hydrocarbons or produced water, but there is no requirement to monitor or report surface leakage emissions. An EOR facility can become a GS facility if they submit a MRV plan and the EPA approves it.

The MRR program leaves gaps at all three levels of GHG accounting. Facility- and inter-facility-level rules are generally adequate for capture and storage facilities, but omit CO_2 transporters. Also absent are any requirements to monitor for leakage in the long-term stewardship phases. The MRR protocols do not provide mechanisms to reconcile the CO_2 inventory tracking reports from the various components of the CCS system and make sure that the technology as a whole is effective. Most importantly, GHG reduction program level protocols are entirely absent: such protocols would integrate the facility and inter-facility reporting to determine the overall effectiveness of CCS and its relative value at producing low-carbon energy.

8.3 TECHNICAL CONSIDERATIONS FOR MONITORING AND MEASUREMENT OF SURFACE LEAKAGE AT GS SITES

GHG accounting for CCS will rely on many types of monitoring and measurements, but the most challenging may be monitoring to measure potential surface leakage at GS sites. The objectives of monitoring for surface leakage are to: 1) ensure that the site is operated in such a way as to protect human health and the environment; 2) demonstrate that CO_2 that is injected is effectively sequestered; and 3) assure the integrity of the GHG emission reduction program. While protection of human health and the environment is the most critical monitoring objective, this is a part of the GS site permitting process rather than the accounting process. Monitoring required for the permitting process will serve as the foundation for GHG accounting, but GHG accounting may impose additional monitoring requirements.

This section discusses monitoring specifically as it relates to measuring and reporting surface leakage. Monitoring for surface leakage involves three primary issues:

- What amount of surface leakage should a monitoring program be able to detect?
- What monitoring techniques and monitoring strategies are most appropriate?
- What is the cost of surface leakage monitoring?

The monitoring issues discussed in this section will be vital to address under a GHG reduction program, but they apply equally to the monitoring required at GS sites under the MRR.

Detection Limits

Perfect retention of injected CO_2, while highly desirable, is not necessary for CCS to be an effective greenhouse gas mitigation technology.[12] Recent research investigating the impact of surface leakage on the overall effectiveness of the CCS system has used the metric of annual leakage rates (i.e., some fraction of the sequestered CO_2 leaks each year) to investigate the climate and economic impacts of leakage. Research suggests that an overall effective annual leakage rate of 0.1% or less of the total mass of injected CO_2 is nearly equivalent to perfect storage.[13] Another recent study argues that CCS could have significant economic and climate-control values even with annual leakage rates at the storage site of 1% per year,[14] depending largely on the relative costs of CCS and alternative low-carbon energy sources. While these studies assessed leakage in terms of percentage of mass injected, if leakage actually occurred it would not necessarily be proportional to the total amount of CO_2 in place, nor did it need to occur continuously over the life of the project.

Operationally, detection limits will need to be set in terms of leakage quantity, not leakage rate, and GS site monitoring plans will need to be designed to measure and report the mass of CO_2 lost through surface leakage. Detection limits based on the theoretical relationship between CCS effectiveness and a certain percentage of leakage must therefore be translated into leakage quantities. A specified percentage leakage rate would produce varying annual emissions depending on the site size and years of operation. For example, at a site injecting an average of 1 Mt CO_2 per year, 0.1% annual leakage would total approximately 1,000 tonnes per year in the first year, and then 30,000 tonnes per year after thirty years of injection. Furthermore, in such an escalating formulation, at a site injecting an average of 10 Mt CO_2 per year, annual leakage of 0.1% would total approximately 10,000 tonnes per year in the first year, and 300,000 tonnes per year after thirty years of injection.

A second dimension that should be considered in setting surface leakage detection limits is the reliability of emissions measurement for other elements of the CCS system. To assure the integrity of a GHG emission cap, surface leakage measurement must be at least as reliable as the measurement methods used for GHG emissions reporting in other CCS processes. One obvious point of comparison is with the meters used to measure CO_2 injection volumes, which have a typical level or error of about ±1%.[15] For GS sites injecting 1 and 10 Mt CO_2 per year, this translates to an ambiguity in the amount of mass injected of 10,000 and 100,000 tonnes, respectively. For pipelines, analogies can be drawn from the gas industry, where realistic expectation for "loss or otherwise unaccounted for" gas in transmission systems is typically ±0.25% to ±0.50% on a volumetric basis.[16] Reliability of emission measurements from CO_2 capture sources is not yet well understood, and may vary by capture process. The MRR established a minimum reporting threshold of 25,000 tonnes per year for stationary combustion sources like electric power plants.

From the discussion above, it is apparent that setting detection limits for surface leakage is a complex decision. Overall, however, it appears that surface leakage detection limits on the order of 1,000 to 25,000 tonnes per year would satisfy criteria of ensuring the effectiveness of CCS and holding GS sites to similar standards as other components of the CCS system. Given the large range of project sizes and the long operating lifetimes of projects, it will likely be necessary to use a more nuanced rule than a simple fixed fraction of the cumulative CO_2 injected at the site. One option would be to make detection limits site specific and modify them over time based on the amount of CO_2 in storage at the site. This is the approach the MRR has taken. Operators are charged with the responsibility to propose "a strategy for detecting and quantifying any surface leakage of CO_2."[17] As discussed in Chapter 5, this type of performance-based approach has advantages in being flexible to accommodate varying sequestration site geologies and operational configurations, although it will require substantial expertise on the part of regulators to implement. Additionally, regulators will also need to coordinate closely in order to promote consistency.

Monitoring Technologies and Monitoring Strategies

Monitoring technology and monitoring strategy are inextricably linked. Technical capabilities will guide selection of an overall monitoring strategy. The monitoring strategy will then form the basis for a site-specific monitoring plan to detect and measure surface leakage. Monitoring methods that could contribute to measurement of surface leakage include techniques designed to:

- measure the mass of CO_2 in situ;
- measure surface leakage across the entire site;
- detect a loss of CO_2 containment from the injection formation;
- detect surface leakage in a targeted area; and
- measure emissions once surface leakage is located.

The monitoring strategy for the site-specific monitoring plan will be determined by the availability of cost-effective monitoring methods to achieve each of these five objectives in an accurate and precise fashion, given the unique geologic and man-made features of the site. Figure 8.1 summarizes the interplay between monitoring technology capabilities and monitoring strategy selection. While future improvements in monitoring methods may raise new possibilities, there is currently only one general monitoring strategy that emerges as feasible: targeted surface monitoring to locate and measure surface leakage *only if* subsurface monitoring indicates CO_2 has escaped confinement and migrated to the surface. This strategy can be used as a basis for a wide variety of monitoring plans, as operators select the combination of monitoring methods that are the best fit for the geologic and man-made characteristics at their site. The following paragraphs describe the current status of monitoring methods to meet each of the five objectives listed above, and also discuss the implications of monitoring capabilities for monitoring strategy.

Measuring the Mass of CO_2 In Situ

Currently, the accuracy of direct or indirect monitoring techniques to measure the subsurface mass of CO_2 in situ is low. For example, estimates of the mass of stored CO_2 derived from seismic studies are at best accurate $\pm$ 20%,[18] so they are not adequate for GHG accounting. This means that calculating surface leakage by subtracting the measured mass of CO_2 in the subsurface from the measured mass of CO_2 that was injected is not a feasible strategy.

FIGURE 8.1 *Decision process for selecting a monitoring strategy to measure surface leakage for GHG accounting. Potential monitoring strategies and site selection indicators that emerge from four key site-specific monitoring questions. "Yes" branches assume that cost-effective monitoring technology is available that will be accurate and precise given the site-specific geologic conditions.*

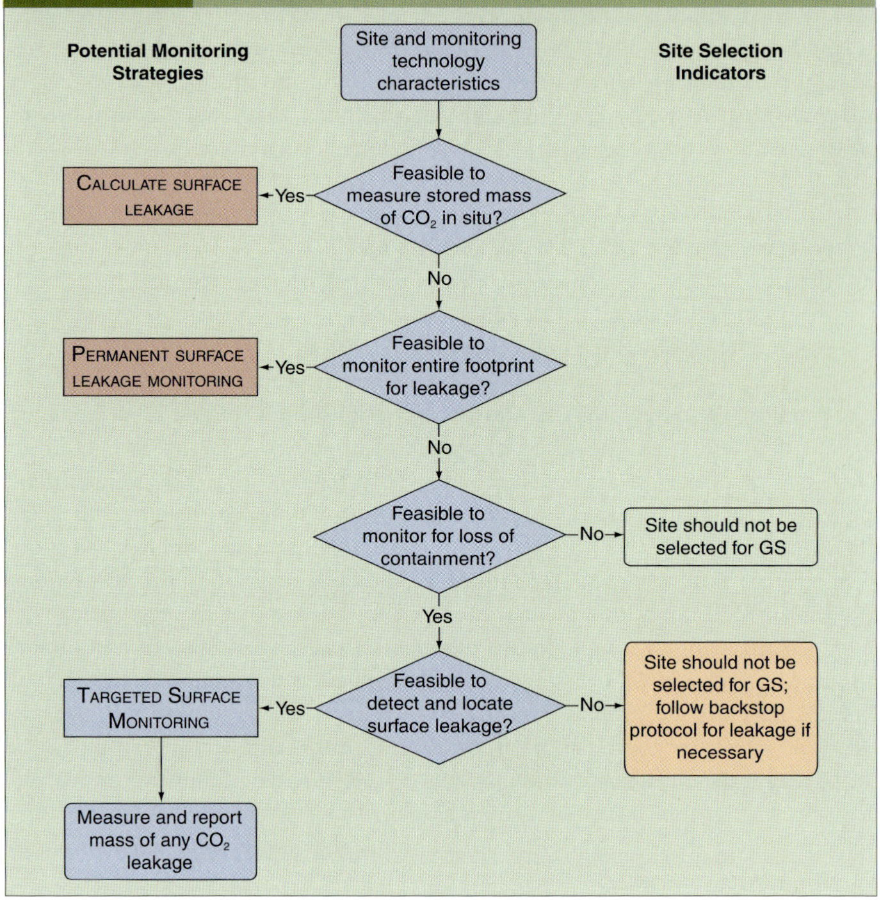

Measuring Surface Leakage Across the Entire Site

Monitoring methods designed to measure surface leakage routinely across the entire site are currently neither sufficiently accurate nor cost effective to form the basis for GHG accounting. Monitoring methods suggested for this purpose include eddy covariance, accumulation chambers, soil gas probes, and laser-based methods.[19] Properties of these monitoring methods and others are described in Table 8.1. Proposals have been made to detect leakage by setting up permanent monitoring on a fixed grid over the site, but researchers increasingly question the effectiveness of this approach.[20] Eddy covariance monitoring would have trouble detecting a leakage signal because CO_2 is rapidly diluted once it passes from the ground into the atmosphere.[21] Additionally, the relatively small signal of elevated atmospheric CO_2 due to surface leakage would be difficult to differentiate from the noisy background of biological and industrial processes at a GS site.[22] Accumulation chambers and soil gas monitoring both measure such a small area (~1 m^2) that a very dense grid would be needed to have a sufficiently high probability of detecting leakage, and field studies suggest that laser techniques are better suited to targeted leakage detection rather than site-wide monitoring.[23]

This means that it is not currently feasible to use the potential monitoring strategy of permanent surface monitoring to conclusively demonstrate there is no surface leakage. Some argue that regardless of the care with which sites are chosen

TABLE 8.1 ***Selected leakage monitoring and measurement techniques.***

Method	Detects	Proposed Use	Measurement Footprint
Visual observation or hyperspectral imaging of vegetative stress	Elevated soil gas CO_2	Leakage detection	Unlimited (where vegetation is present)
Soil gas survey	Elevated soil gas CO_2	Leakage detection and measurement	~1 m^2
Accumulation Chambers (AC)	CO_2 flux from ground to atmosphere	Leakage detection and measurement	10 cm^2
Eddy Covariance (EC)	CO_2 flux in near surface atmosphere	Leakage detection and measurement	10^4 to 10^6 m^2
Laser methods	CO_2 concentration in near surface atmosphere or shallow subsurface	Leakage detection and measurement	10 to 10^3 m linear trace

Source: Adapted from: IPCC. 2006. Carbon dioxide transport, injection and geological storage. In *2006 IPCC Guidelines for National Greenhouse Gas Inventories, Vol. 2.* Eggleston, H. S., L. Buendia, K. Miwa, T. Ngara, and K. Tanabe, eds. Hayama, Japan: IGES.

and operated, there will still be unknown pathways for leakage, and therefore monitoring must conclusively show that leakage is not occurring. The fact that this monitoring strategy is currently unavailable could present a problem if a particular amount of surface leakage would render CCS ineffective, but this is not the case. As discussed above, depending on project and site details, CCS can be an effective climate mitigation technique as long as surface leakage averages in the range of 0.1% to 1% of the mass of CO_2 in storage per year. Research indicates that, if subsurface monitoring indicates that CO_2 has escaped confinement, then there is a high likelihood of locating and measuring this level of surface leakage using targeted surface monitoring to locate and measure surface leakage *only if* subsurface monitoring indicates that CO_2 has escaped confinement.[24]

Detecting a Loss of CO_2 Containment from the Injection Formation

Monitoring to detect loss of confinement is feasible with current technology. There is a broad portfolio of direct and indirect monitoring methods that could give early warning if CO_2 escaped the confining zone.[25] Methods include seismic, gravity, pressure, and geochemical monitoring. In general, the strength of subsurface monitoring is its ability to detect the presence of free-phase CO_2. For example, seismic techniques can detect secondary accumulations of CO_2 as small 1,000 tonnes at 1 km depth, and 100 tonnes at 500 m depth.[26]

If subsurface monitoring were to indicate that CO_2 had leaked past the confining layer, the next step would be targeted monitoring to detect whether CO_2 had migrated all the way to the surface and was leaking to the atmosphere.

Detecting Surface Leakage in a Targeted Area

The likelihood of success in using targeted surface monitoring as a strategy to measure the mass of surface leakage for GHG accounting is based on knowledge of the physical processes that govern CO_2 movement in the subsurface, and on the demonstrated capabilities of currently available monitoring methods. Trapping of oil and gas under deep confining layers over millions of years demonstrates that where site conditions are appropriate, impermeable rock layers can successfully sequester a buoyant fluid like CO_2. Geologic studies also indicate that if CO_2 were to escape confinement, it is highly unlikely to emerge as uniform surface leakage across the entire GS site. Instead, CO_2 would preferentially migrate upward following geologic or man-made pathways, like faults or wells.[27] This has been demonstrated in areas of natural subsurface CO_2 accumulations, such as Mammoth Mountain, California, the Paradox Basin, Utah, and many sites in Italy. The spatial distribution geometry of surface leakage at any particular site would be a function of the geologic and man-made site-specific features that breach confining units and the physical processes that govern CO_2 migration from the injection formation upward through thousands of feet of rock to the surface. Regardless of how

TABLE 8.2 *Plausible scenarios if CO_2 leakage were to occur.*

Final Pathway to Surface	Surface Leakage Footprint	
	Shape	Size (Order of Magnitude)
Well: terminates above water table.	Point	1 m^2
Well: terminates below water table, CO_2 rises through water table aquifer, flows vertically through vadose zone.	Small area	10 m^2 to 100 m^2
Fault or fracture: CO_2 emerges from fault or fracture, possibly spreading if it passes through unconsolidated deposits above the bedrock.	Linear	1,000 m^2 to 1 km^2

CO_2 makes its way out of the injection formation and upward through multiple geologic formations, the size and shape of the leakage footprint will be determined by the nature of the final pathway to the surface. In general, however, only a limited number of plausible scenarios could result in surface leakage, and each would produce a characteristic leakage footprint, ranging in size from approximately 1 m^2 to 1 km^2 as described in Table 8.2.

Monitoring to detect surface leakage could be based on measurements of CO_2 concentrations in the soil near the surface, the flux of CO_2 from the surface to the atmosphere, or the flux (or concentration) of CO_2 in the near surface atmosphere (i.e., 3 to 30 meters above the ground) (see Table 8.1). Even though none of these approaches is fully proven as a means of leakage detection, early field results and experience with natural analogs suggest that screening based on elevated soil gas concentrations is promising.[28]

Measuring Emissions Once Surface Leakage is Located

Once monitoring has identified the area over which leakage is occurring, well-established monitoring methods with decades of proven efficacy in environmental and ecological studies are available to measure the leakage rate and estimate the total amount of CO_2 that is escaping to the atmosphere, including soil gas surveys and accumulation chamber studies.

The success of targeted surface monitoring at any particular site hinges on the availability and cost effectiveness of monitoring methods to detect loss of confinement, then detect, locate, and measure surface leakage at levels above the detection limits occurring over plausibly sized leakage anomalies, given the specific geologic characteristics of that site. Generally, for many geologic conditions, evidence suggests that this will be possible.

Cost of Monitoring for Surface Leakage for GHG Accounting

No currently published studies estimate the cost of targeted monitoring to measure surface leakage. Any such estimate would be inherently limited by the fact that different site conditions will call for very different monitoring methods, and by the fact that the cost to employ them would vary depending on the circumstances of the leakage incident. For example, even at the same site, the cost to find and measure CO_2 leaking to the atmosphere through an abandoned well would be less than the surface leakage monitoring costs if CO_2 were to travel upward along complex, intersecting faults and fractures.

Despite the lack of historic data on monitoring costs, several observations suggest that the cost of monitoring added by the demands of GHG accounting would be small compared with the cost of monitoring that would be required by a UIC Class VI permit to screen for health, safety, or environmental impacts and to establish eligibility for post-injection liability transfer. Because a UIC Class VI permit will require monitoring to detect loss of containment, as well as monitoring to locate surface leakage should loss of containment occur, the only monitoring costs added by the demands of GHG accounting would be those associated with measuring the mass of leakage after the surface leakage is located. This type of monitoring could be accomplished using techniques such as accumulation chamber and soil vapor studies, which are well-established monitoring methods, as described above.

8.4 COMPENSATING FOR SURFACE LEAKAGE

The limited precision with which it is likely that surface leakage can be measured raises questions about the best way for GS site operators to "pay" for such leakage. As discussed in the next section, the form that this payment could take would depend on the nature of the GHG reduction program. For example, under an economy-wide cap and trade program, a GS site operator might buy allowances to compensate for surface leakage, whereas under a program where GS sites sell offsets, surface leakage might be "deducted" from the value of the offsets. The alternative of simply charging a penalty to operators who detect loss of confinement, with no monitoring or measurement of surface leakage required, is not incompatible with the new UIC Class VI rules and the MRR.

The current emission reporting rules for GS under the MRR require GS sites to measure the mass of surface leakage if loss of containment is detected. But for credibility, a GHG reduction program should also include protocols to deal with the uncertainties associated with GS as a nascent technology with the surprises that can arise when relying on natural systems. In other words, what if monitoring technology performs significantly more poorly than expected at a GS site, leaving an operator who knows that CO_2 has escaped the confining zone unable to locate or measure surface leakage? High-quality site selection criteria should guard against this scenario, but having protocols in place to deal with the unexpected and protect the integrity of the GHG reduction program would also enhance the credibility of CCS. The MRR provides for "what if" scenarios in a number of areas by specifying protocols on how to make up for missing data.[29] An emissions reduction program would benefit from including a protocol to regularize uncertainty by specifying how surface leakage should be compensated for in the presence of insufficiently accurate or precise monitoring.

A backstop method for compensating for surface leakage from GS sites could be based on a percentage of either the annual mass of CO_2 injected or the total inventory of CO_2 sequestered. No surface leakage emissions would be reported as long as no loss of containment was detected. However, if routine monitoring detected that CO_2 had escaped the confining unit(s) and the regulator determined that surface leakage emissions could not be quantified, the GS site operator would be required to take compensatory action to cover some percentage of the total amount of CO_2 sequestered.

More research is needed before a backstop compensation level could be established. Clearly it would have to be set high enough that operators would prefer to quantify surface leakage, as the intent is to create an incentive for creation for ongoing improvements to monitoring technology. However, the level should not be set so high (perhaps a default level of less than a few tenths of a percent of injected CO_2) that it would unduly penalize sites that do not appear to be leaking significant amounts of CO_2, but are otherwise unable to quantify surface leakage due to limitations of monitoring technology.

The issue of compensating for surface leakage under a GHG reduction program is further complicated by the transition that will occur when a site ceases operation and enters long-term stewardship. During operations and the post-injection period, operators will have a revenue stream that they can draw upon to "pay" for surface leakage. And while the operator will be directly liable for the site, GHG accounting will be a matter of the national government ensuring that a private party meets its obligations under the GHG emission reduction program.

Compensating for potential surface leakage during long-term stewardship is more complex. As explained in Chapter 7, under the regulatory approach proposed, no site would be accepted for long-term stewardship by the Federal Geologic Sequestration Board unless it was stable and did not show signs of significant leakage. However, it is of course still possible that leakage might occur at such a site at some later time. While most of the environmental and technological implications of leakage would not change when long-term stewardship begins, there are critical practical and institutional differences from the earlier phases. During long-term stewardship, the probability of leakage will be very small. For this reason, some suggest that this justifies having no requirements for monitoring or compensating for surface leakage during long-term stewardship. This approach would have the advantage of simplicity and cost effectiveness and is supported by knowledge of the CO_2 trapping mechanisms in the subsurface[30] and the expected rates of pressure decrease post-injection. However, there are other long-term mechanisms that could offset some of this decrease in the risk of surface leakage (e.g., slow geochemical reactions that increase the permeability of confining formations and abandoned artificial penetrations). Without protocols to handle surface leakage during long-term stewardship, the risk of post-operational surface leakage is transferred to future generations, raising questions of intergenerational equity. Protocols to handle surface leakage during long-term stewardship would improve the integrity both of the GHG reduction program and of CCS as an emission reduction technology.

Institutionally, compensating for leakage during long-term stewardship is more complex than during operations, because no revenue stream will be available. This means that resources to compensate for potential surface leakage during long-term stewardship must be set aside while a GS site is operating. Regulatory options to fund compensatory action for surface leakage during long-term stewardship would vary considerably, depending on the form of the GHG reduction program. This issue is discussed in Section 8.5.

8.5 CCS ACCOUNTING UNDER VARIOUS CLIMATE OR ENERGY POLICIES

Accounting protocols for CCS will vary, depending on the type of climate or energy program(s) put in place to control GHG emissions. Programs that contain a cap on emissions offer the simplest accounting options, and an economy-wide cap covering all three elements of CCS would be simplest of all. Under a cap and trade system, the amount of CO_2 that can be emitted is fixed, and alternative low-carbon energy technologies compete on merit and cost. CO_2 is either emitted or not emitted, program objectives are met as long as emissions are accurately measured and reported, and emitters have allowances to cover their emissions. Programs with no cap that aim to control GHG emissions by encouraging low-carbon energy technologies through portfolio or performance standards can encounter the difficulties discussed in Box 8.1, where program objectives require estimating emissions avoided by using CCS compared with some other type of energy production. This would result in more complicated GHG accounting procedures.

In the paragraphs that follow, we outline how GHG accounting for CCS would function under a cap and trade program, a portfolio standard, and a performance standard by describing the compliance mechanism; noting whether CO_2 generators will report captured CO_2 as emissions; explaining the relationship between the CO_2 generator and the GS operator; describing accounting protocols for GS site emissions during operations; and suggesting options to compensate for surface leakage from GS sites in long-term stewardship. Program characteristics are sketched in simplified form to illustrate major principles, but GHG reduction programs could be designed that would combine and/or modify the program elements described here. For this discussion, we assume the capture and sequestration facilities are owned by separate entities, and we ignore the transportation element of GS. Table 8.3 summarizes GHG

Table 8.3 *GHG accounting for CCS under different types of GHG reduction programs.*[a]

Ghg Reduction Program Type	Compliance Mechanism	Captured CO_2 Reported as Emissions?	Relationship Between CO_2 Generator and GS Operator	Ghg Accounting at GS Sites	Accounting for Surface Leakage During Long-term Stewardship	Notes
Cap & Trade (all parts of CCS are capped)	Binding emission cap. Emitters must submit allowances.	No	Generator pays GS operator to take CO_2.	Submit allowances to cover GS site emissions.[b]	GS operators pay into fund. Long-term steward submits allowances.	Simplest system for GHG accounting for CCS.
Cap & Trade (GS sites not capped)	Binding emission cap. CO_2 generators must submit allowances and/or offsets.	Yes	GS site operator sells offsets to generator.	Offsets = imported CO_2 minus GS site emissions.	GS operators pay into fund. Long-term steward submits allowances.	Systems that use offsets must consider "additionality" and whether emissions are actually reduced or just displaced (emissions leakage).
Portfolio standard	Percentage of energy supplied must come from approved sources. Emitters can buy Clean Energy Certificates (CECs).	Yes	GS site operator sells CECs to generator.	Track avoided emissions of imported CO_2. CECs = avoided emissions minus GS site emissions.	GS operators pay into fund. Long-term steward buys CECs.	Most complicated system for GHG accounting for CCS.

(Continued)

Table 8.3 *(Continued)*

GHG Reduction Program Type	Compliance Mechanism	Captured CO_2 Reported as Emissions?	Relationship Between CO_2 Generator and GS Operator	GHG Accounting at GS Sites	Accounting for Surface Leakage During Long-term Stewardship	Notes
Performance standard	Energy supplied meets maximum emission rate (e.g., lbs CO_2/kWh).	No	Generator pays GS operator to take CO_2. Generator must periodically "true up" their emission rate to account for GS site leakage.	GS site emissions reported back to generator for "true up."	No direct way to link future surface leakage to current compliance obligations.	Accounting protocols less complex than with portfolio standard, but fewer incentives for high quality GS site performance, and difficult to compensate for surface leakage.

Notes: a. This table assumes a disaggregated CCS industry. GHG accounting for a vertically integrated CCS industry would differ somewhat.
b. GS site emissions include surface leakage, equipment leakage, and vented CO_2 as specified by the MRR program.

accounting issues for a disaggregated CCS industry under different types of GHG reduction programs.

Cap and trade, the most familiar market-based GHG control program, works by setting a cap for the emissions from the entire economy or from certain sectors, then allowing emitters to buy and sell allowances. Although much of the academic literature focuses on the "trade" elements of cap and trade, the key aspect of the program's environmental merit lies in the level at which the "cap" is set and how it evolves over subsequent years. "Trade" provisions are simply mechanisms for finding the lowest cost source of emission reductions and are environmentally significant only if they either dilute responsibility for honoring caps or reduce compliance costs to levels that allow more stringent cap-setting.

There is no standard design for a GHG cap and trade program. A "downstream" cap and trade program would allocate or auction a fixed number of tradable allowances to emitters and requires those emitters to surrender allowances equal to their emissions in a particular compliance period. An "upstream" program, however, would require firms to surrender allowances equal to the carbon content of the fuel they distribute in commerce each year. A regulatory program can combine both upstream and downstream elements.

The choice of an "upstream "or "downstream" program has important regulatory implications for the deployment of CCS. Under a downstream system, emitters (such as power plants and large industrial sources) are responsible for submitting allowances to cover their emissions. If a facility captures CO_2 and exports it for permanent sequestration, then no emission occurs and no allowance submission should be necessary, as long as regulators are satisfied that the CO_2 will not leak back into the atmosphere.

In an upstream system, accounting for sequestration is more complicated. Here, the point of regulation is not the emitter, but rather a producer, processor, or distributor of the fuel that the emitter combusts. The producer, processor, or distributor is required to submit allowances to cover the carbon content of that fuel. The cost of allowances is built into the price of fuel sold to the downstream emitter. However, because the emitter has no allowance submission requirement, it has no incentive to sequester unless some form of sequestration credit is provided. For that reason, a well-designed upstream cap and trade program would provide a sequestration credit to downstream entities that sequester CO_2 produced from fuel that was subject to an upstream allowance submission requirement. Several recent US legislative proposals combine elements of upstream and downstream regulation.

Cap and trade systems also vary in how allowances are initially distributed, and in the proportion of compliance obligations that can be met using offsets from emission reduction activities that are outside the boundary of the cap. The conventional wisdom is that a market-based control system works only if regulators and market participants are confident that the "regulatory currency" (i.e., allowances) will not be debased by under-reporting of emissions or over-reporting of offsets. This highlights the importance of well-designed GHG accounting protocols.

During the design of a market-based GHG emissions reduction, two choices must be made about how the components of a CCS project are treated under the program. First, should a CCS project component be considered an emission source (i.e., covered entity) like an electric power plant, or should it be excluded from GHG emission limits, similar to forestry projects? Second, if it is classified as an emission source, should there be an emissions threshold below which emissions reporting is not required?

For a capture facility, the answers to the above questions are straightforward. An industrial facility such as an electric power generator or gas processing plant that captures CO_2 would, by definition, be considered an emission source. To improve economic efficiency of the GHG emission reduction program, facilities emitting less CO_2 than a predetermined threshold should not be expected to report their emissions. Similarly, pipeline systems that transport GHGs such as CO_2 or natural gas and—in the case of most natural gas and some CO_2 pipelines—burn fossil fuels to provide energy for the system would also be included in the regime as an emissions source, subject to reporting thresholds. Many US cap and trade proposals only subject industrial facilities to the cap if they meet certain emissions thresholds (e.g., 25,000 tonnes/year).[31] Applying such thresholds might be appropriate for capture facilities and transport infrastructure, but would create perverse incentives for GS projects. Exempting smaller GS projects from the emissions cap could discourage accurate monitoring and reporting of leakage emissions or incentivize development of smaller individual facilities to avoid reporting requirements. The MRR program has adopted this approach, requiring all GS sites to measure and report their GHG emissions.

Economy-wide Cap and Trade

If GS projects were classified as emission sources under a cap and trade GHG emission reduction program, then CO_2 that is captured and sequestered would be considered to not be emitted. Each of the three components of the CCS value chain—capture, transport, and sequestration—would be treated as a separate facility, with obligations to report all emissions, including energy use, fugitive or vented emissions, and leakage emissions. A GS operator would purchase allowances to cover GS site emissions, including any measured surface leakage (or, if measurement is not possible, use the backstop protocol described in Section 8.4). Allowances to cover potential surface leakage during long-term stewardship could come from a long-term stewardship fund that all GS operators contribute to such as the Carbon Sequestration Trust Fund described in Chapter 7.

Limited Cap and Trade

A cap and trade program might be designed to limit emissions from major stationary sources like electricity generators without establishing compliance obligations for GS sites (or CO_2 transporters). In this case, CO_2 generators would report captured

CO_2 as emissions. Injecting the CO_2 in a GS project would create offsets, which could be sold to the entity that generated the CO_2, or to others in need of offsets, to meet their obligations under the GHG emission reduction program.[32] The quantity of offsets a GS site could sell should be set as the mass of CO_2 injected minus CO_2 lost to venting, equipment leakage, and surface leakage. The potential for surface leakage during long-term stewardship could be handled in the same manner as under an economy-wide cap and trade program, with GS site operators paying into a fund that the long-term steward could use to purchase allowances in the event of surface leakage.

Portfolio Standard

A portfolio standard requires energy producers to include a certain percentage of energy created by preferred technologies in their total energy sales. Emissions are not capped; rather, the goal is to avoid emissions by using low-carbon energy technologies. The most familiar form is the renewable portfolio standard for electricity generation, but a number of states have introduced variants of this type of program designed to include other low-carbon electricity technologies such as CCS or nuclear power (often referred to as clean energy portfolios standards, or alternative energy portfolio standards) as in a number of other states. Portfolio standards applied to liquid fuels, such as California's low carbon fuel standard, could also be a driver for CCS.

With portfolio standards, energy producers may own the low carbon energy production facilities, or they may purchase certificates or credits to meet their obligations. In this respect, portfolio standards include a market element, despite the fact that emissions are not capped. The market for credits provides feedback on the relative cost and emission reduction potential of various emission reduction options.

If CCS is included in a portfolio standard alongside renewable technologies like wind or solar, the accounting protocols should provide mechanisms to compare their relative emission-reducing merits. In other words, GHG accounting will need to estimate avoided emissions, and one tonne of CO_2 sequestered would not be worth a one-tonne credit. The credit would be worth a fraction of a tonne, decreased to represent the actual emissions that were avoided in sequestering the CO_2. This would require significantly more complex accounting systems due to the fact that GS sites could sequester CO_2 from a number of sources (with varying avoided emissions), and could sell credits to entities other than the firms that generated the CO_2. The quantity of credits a GS site could sell should be set as the mass of avoided emissions injected minus the mass of any GS site emissions. This would require accounting protocols that would define how to calculate avoided emissions, and then track avoided emissions a fraction of the total amount of CO_2 as it is transferred from capture to transport to sequestration facility, as part of the CO_2 inventory tracking system.

The potential for surface leakage during long-term stewardship could be handled in a similar manner as under a cap and trade program, with GS site operators paying into a fund that the long-term steward could use to purchase credits in the event of surface leakage.

Performance Standards

Compliance obligations under a performance standard are typically set at a maximum allowable rate of emissions per unit of energy (e.g., tonnes of CO_2 per megawatt-hour of electricity generated or sold). The allowable rate could be set by statute. For example, laws in Washington state and California require GHG emissions from new generation sources to be no more than those from a natural gas power plant, on a per-kilowatt basis.[33] Alternatively, it could be determined on a case-by-case basis, such as Best Available Control Technology (BACT) analyses for GHG emissions under the Clean Air Act. Performance standards could be designed to apply to individual units, to a collection of generators, or to entities that sell (rather than generate) electric power.[34] Under a performance standard, a generator's emission rate would be calculated as the emissions from generating electricity and capturing CO_2 minus the mass of CO_2 exported for sequestration—captured CO_2 would not be reported as emitted. Performance standards differ from portfolio standards in that there is no market element—that is, no trade in credits or offsets. On the one hand, this simplifies GHG accounting. The emission rate is calculated with captured CO_2 treated as not emitted, and no complicated protocols are needed to define and track avoided emissions. On the other hand, a performance standard decreases incentives for high-quality GS site performance and makes mechanisms to compensate for surface leakage cumbersome during operations and nearly unmanageable during long-term stewardship. The decreased accountability for GS site performance comes about because, unlike systems where the GS operator sells offsets or credits, where CO_2 lost to leakage represents lost revenue, with a performance standard it is difficult to link the performance of the sequestration site back to the generator's compliance obligations. During operations, some sort of periodic "true up" could be required, wherein any surface leakage from GS sites would be factored back into a corrected emission rate. Potential surface leakage during long-term stewardship presents greater difficulty in that there is no direct way to link future GS site performance to current maximum emission rates.

8.6 RECOMMENDATIONS ON GHG ACCOUNTING FOR CCS

Our recommendations on GHG accounting for CCS are organized into three sections. Most are applicable under any type of GHG reduction program. For these overall recommendations, instead of referring to program-specific mechanisms to compensate for surface leakage (such as allowances or offsets), we use the generic phrase "compensatory action." Details can be found in the draft legislation in the Appendix. At the end of this section, we discuss GHG accounting protocols for CCS under an economy-wide cap and trade program, and finally, accounting principles that should be considered if GHG reduction programs other than an economy-wide cap and trade program were to be instituted.

Only CO_2 injected at a site with an approved plan to monitor, measure, and report any surface leakage should be considered geologically sequestered under any federal GHG reduction program. Giving credit for CO_2 that is injected without verifying that it is permanently sequestered would undermine the integrity of both CCS and the GHG reduction program. Therefore, CO_2 injected for EOR at sites that have not elected to put in place an approved monitoring plan should be counted as emitted.

Routine permanent surface monitoring to prove no leakage is occurring should not be required. Because routine subsurface monitoring required by permit conditions will include methods that can detect loss of containment, we believe that surface monitoring proving that leakage is not occurring should not be required. The scientific and technical evidence supports this approach in that leakage is unlikely at well-characterized, well-operated sites: GS is beneficial even with low levels of leakage; subsurface monitoring methods to detect loss of confinement are feasible; and focused surface monitoring methods can feasibly locate and quantify leakage of quantities that matter (over 1,000 tonnes per year) through plausibly sized leakage footprints (10 m^2 to 1 km^2).

If a loss of containment is detected through routine monitoring, GS site operators should undertake a targeted monitoring program to measure surface leakage emissions from the site. We recommend that regulation require targeted surface monitoring to measure surface leakage only if subsurface monitoring indicates CO_2 has migrated through the confining formation and surface monitoring detects leakage. The surface monitoring program should be designed to quantify, at a minimum, the detection limits. If a satisfactory monitoring program cannot be implemented at the site (due to technological or other factors), the operator must report surface leakage using the backstop protocol.

Regulators should set detection limits for surface leakage. If subsurface monitoring detects loss of containment, operators will need to know the performance standard for detecting, locating, and measuring surface leakage. This will allow operators to design site-specific monitoring programs, select appropriate monitoring technology, and determine the scope of their monitoring programs. Detection limits should be based on monitoring objectives and technical feasibility. They must also take into account the practicality of implementation, cost effectiveness, and fairness, and create incentives for high-quality GS site operation and monitoring. More research and experience with surface monitoring methods is needed before specific performance goals should be established for widespread commercial GS.

Regulators should develop a backstop protocol to compensate for surface leakage in cases where monitoring and measurement are technically infeasible or inaccurate. The results of our technical and policy analyses suggest that if a loss of containment is detected, current monitoring technology should be able to detect and locate leakage at levels of concern at some but not all sites. The disparities in leakage detection capability produced by geologic and ecosystem variations require a backstop protocol for GS in the near-term. The backstop protocol should be set as a percentage of the total inventory of CO_2 sequestered (probably less than a few tenths of a percent of injected CO_2). The operator would owe the backstop compensation from the last time period he or she could demonstrate containment until the leakage was remedied.

An integrated CO_2 inventory tracking system should be established. A national CO_2 inventory tracking system should be established that collects all reports of inter-facility CO_2 transfers, including the mass of CO_2 captured, exported, imported, and injected. The agency should reconcile these reports annually to verify that carbon dioxide captured at a power plant or industrial facility actually reaches an appropriate geologic sequestration site and is injected at that site for permanent sequestration. The agency should determine the parties responsible for CO_2 lost or unaccounted for (above a threshold quality), and publish estimates of the mass of CO_2 captured, transported, injected, vented, and reemitted as surface leakage, as well as the mass of any CO_2 lost or unaccounted for in handling.

Parties responsible for any CO_2 lost or unaccounted for in handling should be required to take appropriate compensatory action. This is necessary to ensure the effectiveness of CCS as a GHG reduction option and to protect the integrity of the GHG reduction program.

We turn now to recommendations for accounting under a cap and trade program.

Under cap and trade, GS sites should be classified as potential emission sources, regardless of size or emissions level, and injected CO_2 should not be counted as emissions. We recommend that GS projects—along with CO_2 capture facilities and pipelines—be classified as emissions sources under a cap and trade program. This approach cleanly defines the boundaries between the three stages of the CCS process (likely run by different commercial entities) and clearly accounts for GHG emissions from energy used to capture, compress, transport, and inject CO_2. Including the GS project under the regime also protects the integrity of a carbon cap by aligning GHG accounting requirements with project operators' other regulatory obligations to prevent and remediate leakage.

Many US cap and trade proposals only subject industrial facilities to the cap if they meet certain emissions thresholds (e.g., 25,000 tonnes/year).[35] Applying such thresholds might be appropriate for capture facilities and transport infrastructure, but would create perverse incentives for GS projects. Exempting smaller GS projects from the emissions cap could discourage accurate monitoring and reporting of leakage emissions or incentivize development of smaller individual facilities to avoid reporting requirements, which will result in unnecessary and inefficient redundancies. Thus we believe all GS projects should be treated as emissions sources, regardless of the amount of their potential annual emissions—the only exception being for research and demonstration projects.[36]

Under cap and trade, GS site operators should contribute to a fund to address potential surface leakage during long-term stewardship. The potential for leakage emissions during long-term stewardship is low, but mechanisms to account for it are important for building public support for CCS technology and supporting the nation's emission reduction goals. The amount that GS site operators pay into the Carbon Sequestration Trust Fund, managed by the Federal Geologic Sequestration Board,[37] should include an amount dedicated to covering the cost of compensating for any surface leakage during long-term stewardship.

A few principles for GHG accounting for CCS under alternate GHG reduction programs.

If a *cap and trade* program is created that does not treat GS sites as emission sources, then:

- captured CO_2 should be counted as emissions;
- offsets sold by GS operators should be discounted by the amount of emissions from the GS site (i.e., the offset equals injected CO_2 minus GS site emissions);
- GS operators should pay into a long-term stewardship fund. The fund would buy allowances to cover surface leakage during long-term stewardship, if needed.

If a *portfolio* standard is created that includes CCS, then:

- Captured CO_2 should be counted as emissions.
- CO_2 inventory tracking systems should include data on the avoided emission fraction of each tonne of CO_2.
- Certificates or credits sold by GS operators should be discounted to reflect only the emissions avoided by CCS and further discounted by GS site emissions.
- GS operators should pay into a long-term stewardship fund. If needed, the fund would buy credits to cover surface leakage during long-term stewardship.

If a *performance* standard is created that includes CCS, then:

- Care must be taken to assure that CCS does not displace lower-emission energy options.
- The diminished accountability for surface leakage at GS sites should be countered by rigorous attention to permit compliance.

9

MAKING CCS A REALITY

In a market economy like the US, the demand for a product is critical to its adoption and widespread use. In the case of an environmental technology such as CCS, whose sole purpose is to prevent an environmental release to the atmosphere, few if any markets exist in the absence of government actions and policies. For example, why would an electric utility company spend hundreds of millions of dollars to install CCS, even an improved lower-cost capture process, if there is no requirement or incentive to reduce CO_2 emissions? A costly action such as this provides little or no economic value to the company—indeed, the added cost and energy penalty of CO_2 capture increase the cost of operation. Only if a government action either requires CCS, or makes it financially worthwhile to reduce CO_2 emissions, would a sizeable market be created for technologies that enable such reductions. Thus, as with other environmental emissions that affect the public welfare, government actions are needed to create or enhance markets for CO_2-reduction technologies like CCS.

Different policy measures influence markets in different ways. Measures such as government loan guarantees, tax credits, direct financial subsidies, and RD&D funding can help create markets by providing incentives for technology development, deployment, and diffusion. Voluntary incentives such as these are commonly referred to as "technology policy" measures.[1] In contrast, regulatory policies such as an emissions cap, emissions tax, or performance standards that limit emissions to specified levels are mandatory, not voluntary. These policies can create or expand markets for environmental technologies by imposing requirements that can be met only, or at least most economically met, by the use of a low-emission technology like CCS. Decisions about to what to build, when to build, and how to operate a plant with CCS will be based on existing and expected regulations and other public policies.[2] Regulations will determine the availability of funds for capital investment, the viability of the market, and the cost of operations. Regulatory and other policies will define the allowable financial risk-mitigation opportunities (e.g., tax and depreciation schedules, loan guarantees, bonds, and long-term federal stewardship).

At the same time, new regulations, rules, and incentives affecting the construction, operation, and retirement of CCS infrastructure must account for how these facilities will be operated and anticipate the effects they will have on other markets and investment decisions. Indeed, as discussed later in this chapter, without appreciating how an owner will likely operate a CCS system within the regulatory framework, there is a potential for unintended consequences that cause more harm than good.

Previous chapters have discussed many of the fundamental roadblocks to near- and long-term sequestration commitments. They include the lack of any regulatory requirement to control emissions, geologic uncertainties, regulatory uncertainties, liability issues (especially those arising after an injection site has been closed), legal access to pore space, and commercial financing requirements.

The combined effect of all these uncertainties is that they render commitments to sequester open-ended economic exposures that today are likely impossible to finance. Gasification projects capable of capturing CO_2 at low cost are capital intensive and use advanced technologies. They require substantial equity and debt commitments and require careful financial structuring (and potentially state or federal incentives or credit assistance) to attract capital. Projects that commit to unquantifiable sequestration cost risks will simply not be financeable. Until enough is learned about geologic realities, regulatory requirements, and liabilities, large-scale sequestration commitments cannot be made by billion dollar projects needing commercial financing.

9.1 OPPORTUNITIES TO MITIGATE FINANCIAL RISKS

An electric power generation plant with CCS capability is an integrated system that includes a variety of technologies each with its own risks and risk-mitigation opportunities. For example, solutions that would reduce risks for CO_2 pipeline construction may not directly reduce concerns about access to pore space or ownership of geologic formations. Discussed below are several opportunities to mitigation the financial or other risks associated with one or more components of a CCS system.

Strategies to Support Large-scale Demonstration Projects

The construction and operation of multiple full-scale demonstrations of carbon capture and sequestration, especially at coal-based power plants, has been widely recognized as an essential step in assuring reliable cost-effective operation and gaining public and private-sector confidence in the ability of CCS to safely and reliably reduce carbon emissions. In the US, federal government support for such projects has now materialized in the form of cost-sharing programs with the private sector. As a result, nine large-scale demonstration projects—six at power plants and three at other industrial processes—were launched in 2009 and 2010, as summarized in Table 9.1. These projects are scheduled to come online beginning in 2013 (for industrial sites) and 2014 (for power plants). Successful projects of this kind can play an important role in making CCS a commercial reality.

A Federal Sequestration Tax Credit and Investment Tax Credit for CO_2 Pipelines

Currently, 15% of the costs incurred in enhanced oil recovery are eligible for the enhanced oil recovery federal tax credit (claimed on IRS Form 8830). The Energy Tax Incentives Act of 2005 provides a 30% business investment credit for

TABLE 9.1 *CCS demonstration projects supported by the US Department of Energy.*

Performer	Location	Capture Technology	Capture Rate (m tons/yr)	Target Formation	Start Date
PC Power Plants					
NRG Energy	Thompsons, TX	Amine	~0.5	EOR	2015
FutureGen Alliance	Meredosia, IL	Oxy	1.0	EOR/Saline	2015
IGCC Power Plants					
Summit Texas Clean Energy	Odessa, TX	Selexol	3.0	EOR	2014
Southern Company	Kemper County, MS	Selexol	2.0	EOR	2014
Hydrogen Energy California	Kern County, CA	Rectisol	2.0	EOR/Saline	2016
Industrial Processes					
Leucadia Energy Lake Charles	Lake Charles, LA	Rectisol	4.0	EOR	2014
Air Products	Port Arthur, TX	Amine	1.0	EOR	2013
Archer Daniels Midland	Decatur, IL	Amine	1.0	Saline	2014

Source: DOE, Office of Fossil Energy, Washington, DC.

solar energy and fuel cell property and certain solar lighting systems; a 10% investment tax credit is provided for microturbines (claimed on IRS Form 3468). A tax credit in the range of 10% to 30% of incurred costs for carbon dioxide pipelines would be in accord with the federal tax credits used to encourage the above investments.

A sequestration tax credit for geologic sequestration would likely provide an effective incentive for early sequestration projects. Such tax credit should include provisions that reduce the amount when the US enacts legislation resulting in a carbon price above the effective price established by the tax credit. Like the production tax credit, the sequestration tax credit could be designed with time limits for both the date by which the projects must be underway and the termination date for the credit.

The sequestration tax credit might also be designed with a limit to the total that is available, with projects competing on the basis of cost or on the basis of first come, first served. However, care should be taken to assure that such a provision does not introduce uncertainty that might inhibit investment.

A per kilowatt-hour production tax credit, the Renewable Electricity Production Credit (REPC), is currently applied to electricity generated from low-carbon "qualified energy resources" at a "qualified facility."[3] Enacted as part of the Energy Policy Act of 1992, the credit expired at the end of 2001, and was subsequently extended in March 2002 as part of the Job Creation and Worker Assistance Act of 2002.[4] The tax credit then expired at the end of 2003 and was not renewed until October 4, 2004, as part of H.R. 1308, the Working Families Tax Relief Act of 2004, which extended the credit through December 31, 2005. The Energy Policy Act of 2005[5] modified the credit and extended it through December 31, 2007. In December 2006, Section 207 of the Tax Relief and Health Care Act of 2006[6] extended the tax credit for another year, through December 31, 2008. The recent Wall Street crisis and the $700 billion "rescue" bill by the federal government included a renewal of the tax credits.

Section 710 of the American Jobs Creation Act of 2004[7] expanded REPC to include, among other additional eligible resources, refined coal. Refined coal is defined as "a liquid, gaseous, or synthetic fuel produced from coal (including lignite) or high carbon fly ash, including such fuel used as a feedstock."[8] In addition, refined coal is considered a "qualified emission reduction" under the REPC if reductions of at least 20% of the SO_2 and either NO_x or mercury emissions are achieved through burning refined coal, as compared to burning the feedstock coal.[9] The Energy Policy Act of 2005 further expanded the credit to certain hydropower facilities and Indian coal. Indian coal is coal produced from reserves which, on June 14, 2005, were either owned by an Indian tribe, or held in trust by the US for the benefit of an Indian tribe or its members.[10]

Given that the current iteration of the REPC already creates set-asides for two specific coal resources, it would be reasonable to expand the scope of REPC to include coal-fired power plants equipped with CCS as a low-carbon "qualified facility" or a "qualified emission reduction" rather than enact new legislation for a sequestration-specific production tax credit. Including sequestration activities under an already-existing production credit provision of the tax code would decrease the amount of risk associated with learning and uncertainty that typically accompanies the application of economic incentives for new projects.

Advocates of the REPC have cited carbon dioxide control as a motivation, along with reduction of SO_2, NO_x, and mercury. Current prices for SO_2 and NO_x allowances and estimated prices for Hg control sum to 0.9 cent per kWh for an average coal-fired power plant in the current fleet. The REPC provides a tax credit of 1.5 cents per kWh, in 1993 dollars adjusted annually for inflation, for the sale of electricity produced from qualified energy resources at a qualified facility.[11] Currently, the REPC for these technologies is 1.9 cents per kWh.[12] Using the average US electric power industry CO_2 emission rate of 0.62 tonnes per MWh,[13] the present federal production tax credit equates to $16 per tonne of avoided CO_2, when credits for avoided SO_2, NO_x, and mercury emissions are accounted for.

While tax credits offer an important policy lever to jump start the CCS industry, because a typical commercial CCS project will cost over a billion dollars, it is equally important that these credits have stability over time. A situation in which they cycle on and off from year to year in the face of Congressional legislative vagaries will not induce the large investments that will be needed for CCS on a commercial scale.

Financing of Pipelines with Tax Exempt Bonds

Because of market uncertainty, there may be a need to provide some risk reduction for investors in CO_2 pipelines. This is despite the relatively low technical risk associated with pipeline design, construction, and operation.

Tax-exempt financing can be used by private entities under what are called Private Activity Bonds. Qualified Private Activity Bonds are tax-exempt bonds issued by a state or local government, the proceeds of which are used for a defined qualified purpose by an entity other than the government issuing the bonds. Private Activity Bonds can reduce financing costs through lower borrowing rates because the interest paid to bondholders is not includable in their gross income for federal income tax purposes.

Financing with tax-exempt bonds requires strict compliance with a series of requirements and limitations established by the Internal Revenue Code. Many types of projects that are eligible for tax-exempt financing are subject to a federally required annual volume cap, which restricts the amount of tax-exempt Private Activity Bonds that can be sold in any one state. Starting with 2007, the volume cap each state receives equals $85 per capita per year.

There are two ways tax-exempt financing could be made available for CO_2 sequestration infrastructure investments, which could be defined to include investments in compression, pipelines, and sequestration facilities. One is for the IRS code to be amended to identify specifically CO_2 sequestration investments as "Qualified Private Activity Bonds." Qualifying the investments would enable Private Activity Bonds to be issued under existing IRS rules and would be subject to state volume-cap allocations. Alternatively, legislation could be passed to create a separate allocation of Private Activity Bond authority for CO_2 sequestration investments. This type of legislation was passed to provide GOZONE bonds for states affected by Hurricane Katrina and for the Liberty Zone around ground zero in New York City. Since demand for tax-exempt bonds often exceeds volume-cap restrictions, it is probably preferable to create a separate allocation and program for CO_2 sequestration investments in this manner.

To create a separate bond allocation for CO_2 sequestration investments, federal legislation would be needed similar to the GOZONE legislation. A national cap on the amount available (say $20 billion) would be established and bonding authorities from states with potential projects could apply for volume-cap distributions under the program. One complication that the program would need to overcome is that Private Activity Bonds must be issued by local government entities, making it somewhat unclear how such bonds might be issued for pipelines that might traverse several counties or states.

Direct Federal Payments Per Tonne of Sequestered CO_2

The Department of Energy has undertaken a number of Regional Carbon Sequestration Partnerships[14] to inject CO_2. Some of these, like the Midwest partnership in the Illinois basin, have begun EOR injections as the second phase of the regional partnerships. Phase III regional partnerships have recently had their schedule advanced and the project sizes enlarged. Total Phase III injections will be approximately 21 Mt (seven projects, each of 1 Mt per year for three years).

It is possible that some of the commercial gasification projects currently in advanced planning may be suitable for these Phase III projects (although most of the Phase III projects have already identified their CO_2 sources). The CO_2 stream from a single commercial gasification project represents at least three times the amount of CO_2 envisioned for any of the Phase III sites. There may not be a match in the timeframe of planned commercial gasification facilities with the newly advanced timetable for the Phase III regional partnerships. The limited time (three years) of the Phase III projects is unlikely to be attractive to investors.

A better match would be a program that supported about ten commercial-scale plants with direct or indirect payments to each for sequestration. Such a program would rapidly advance commercial-scale experience. CCS projects currently being planned are said by developers to be able to transport and sequester at $20–60 million per plant, so the total costs would be in the range of $200–600 million annually (the current tax expenditures to support the wind portion of the federal production tax credit were approximately $2.1 billion in 2010). Further analysis is needed to refine these cost estimates.

Such a program might be financed by direct federal appropriation or by a voluntary or mandatory fee on new coal-based facilities. For example, such a facility might pay $5 per tonne of CO_2 into a fund that would be matched by a like amount from federal funds. The sequestration funds would be spent on a limited number (say ten) sequestration projects for at most ten years. A variation on this program would involve the establishment of federal CO_2 sequestration sites for non-EOR sequestration before a carbon price makes private sites profitable. In addition, a system of "bonus" allowances could be implemented where one tonne of sequestered CO_2 earns credit for more than one tonne. See Chapter 10 for details of this risk-reduction mechanism.

State PUC Actions

A finding by the public utility commission of a state in which a CCS facility is located that carbon capture, transport, and sequestration charges are just and reasonable would allow cost recovery for the incremental costs of CCS through the rate base. Similarly, a CO_2 pipeline might be entered into the rate base if it is found to be just and reasonable, and if the pipeline is used and useful.

As with recent legislation in Indiana, "no look-back" provisions would be required to ensure that investors have adequate assurance that the terms of such an approval would not be changed after plant or pipeline construction.[15] Indiana has agreed to a covenant that explicitly provides that "neither the commission, nor any other state agency, political subdivision, or governmental unit may take any action" that has the "effect of limiting, altering, or impairing a utility's right to recover costs" in connection with, or resulting from, a contract to purchase substitute natural gas.[16] Specifically, if the Indiana Utility Regulatory Commission (the commission) approves a utility contract for the purchase of substitute natural gas, or electricity generated in connection with the production of a substitute natural gas, the commission must allow the utility to recover, on a timely basis throughout the duration of the contract, all costs incurred under a contract to purchase of substitute natural gas, as well as all related costs for generation, transmission, transportation, and sequestration.[17] Moreover, the commission is prohibited from taking any action during the contract term that would adversely affect a utility's right to timely recover costs, regardless of any changes in market conditions or other similar circumstances.[18]

State Electricity and Natural Gas Excise-tax Forgiveness or Tax Credit for Energy Required for CO_2 Pipelines and Underground Injection

Many states levy an excise tax on the sale of electric power. For example, Pennsylvania charges a Gross Receipts Tax that at present nets out at 6.45% on all power sold to an end-use consumer within the Commonwealth (wholesale transactions between generators and load-serving entities are not subject to the tax).[19] Electricity generated in Pennsylvania and sold to another state is subject to the Gross Receipts Tax if a similar tax is imposed by that state on power generated in that state and sold into Pennsylvania.[20] Maryland's Gross Receipts Tax is 2%. These taxes generally range from 2% to 4%, although some are as low as 0.6%. Ohio does not have a Gross Receipts Tax but does have a tax on the consumption of electricity.

States that wish to encourage investment in transport and sequestration of CO_2 could exempt from this tax those uses of electric power directly relating to pumping and injection.

A CCS Trust Fund

A proposal developed for the Pew Center on Global Climate Change[21] envisions a generation fee on coal-fired electric power generators. This fee would go to a trust fund for carbon-capture and sequestration projects. A fee of 0.04–0.05 cents per kWh would support a program of $7–10 billion per year, providing the incremental CCS costs for both new plants and retrofits of existing plants in a ten-plant test program. If the fee were 0.11 to 0.14 cents per kWh, the trust fund would support a thirty-plant program. The fees would be lower if plants provide project cost-sharing. Uncertainties associated with this proposal include who should administer the trust fund, and whether including the fees in customer bills requires approval by each state's public utility commission or can be authorized by federal legislation.

Such federal legislation was introduced[22] in June 2008 as H.R. 6258 by US Representative Rick Boucher (D-VA), Chairman of the House Energy and Air Quality Subcommittee, and fourteen co-sponsors. This measure has persisted in more recent House bills, although it seems unlikely that such a measure will be adopted in the near future.

An alternative manner of implementing such a fee may be a proactive industry initiative, similar to the 1988 "Pork. The Other White Meat.®" marketing fee, in which federal legislation enabled a two-thirds vote of pork producers to levy a mandatory assessment on all producers. To date, the leadership of the US electricity industry has been reluctant to support such an approach.

The Pew study authors point out that the use of an independent or quasi-public trust fund entity would both ensure private-sector contracting and staffing standards and avoid the annual federal appropriations process. For example, the programmatic aspects of the fund should be managed by an independent, non-profit organization under contract with the federal government and selected through a competitive bid process. Additionally, the government should utilize the competitive bid process to select a private bank to handle the accounting, investment, and distribution of fund assets. Optimally, the bid contract timeframe should be either for five years with a five-year option to extend, or for ten years with a five-year break-off point. The option to extend emphasizes the need for contingency, whereas the use of a break-off point emphasizes the importance of program stability and commitment.

Congressional approval would be necessary to authorize the creation of such a CCS trust fund. However, because the existence of a congressionally approved fund does not alone generate money (i.e., it merely receives and distributes the fund assets), federal legislation detailing how, from whom, and where the funds will be generated is also needed. As noted in the Pew proposal, clear objectives must be established and it must be reiterated that the fees from the trust fund to the project would terminate when the objectives are reached. Therefore, federal legislation must also establish general project mandates, create specific criteria for measuring project performance, and layout at least a moderate-level detail regarding implementation—too little detail would provide no guidance, but too much detail could lead to unproductive earmarking.

The Pew study authors note that there is a continuum of trust funds, including purely federally administered ones like the highway trust fund (supported by fuel taxes), funds managed by a consortium of stakeholders such as the Ultra-Deepwater and Unconventional Natural Gas and Other Petroleum Resources Fund (administered under Department of Energy oversight), and the Tobacco Master Settlement Agreement (dispersed by the National Association of Attorneys General).

Such a program could provide incentives for CCS plants using a number of different technologies in various geographic regions, with sequestration in geologic formations of different types.

Carbon Sequestration Investment Fund

Similar to the CCS Trust Fund idea, a program could be developed that would enable projects to agree to make a certain level of sequestration investment as part of their qualification for other government incentive programs. This concept is aimed at providing a mechanism for projects to make firm commitments to CO_2 sequestration initiatives (funding commitments), but not be forced to make commitments to sequester certain tonnage amounts since the technological success and costs of geologic sequestration remain unknown. Under this type of mechanism, only projects that capture CO_2 could qualify. These projects would develop sequestration plans and commitment to invest a certain amount in geologic sequestration activities. For example, for every tonne of CO_2 that they produce, they could agree to spend \$3 per tonne, or \$12 million per year for a 4 Mt plant, on geologic sequestration activities. The commitment would be the basis for the projects qualifying for tax incentives, loan guarantees, or other incentives available for gasification (or other coal technology) projects.

A Carbon Sequestration Investment Fund would be a more flexible (and cost-effective) alternative to the sort of performance standards favored by some. Like other voluntary programs implemented prior to regulations, this investment fund may greatly reduce the costs of technologies prior to their economy-wide implementation.

A Carbon Sequestration Registry

If CO_2 sequestered in the near-term was accounted for by a respected party, and leakage was similarly monitored or appropriately estimated, such a registry might enable a private entity to buy and sequester carbon dioxide in the expectation that they could monetize the sequestered CO_2 when the US creates a regulatory environment that supports carbon dioxide trading. Such a carbon venture fund would be expected to lobby to ensure that pre-existing sequestration projects were counted in legislation that enables any such regime, perhaps as offsets.

The Voluntary Reporting of Greenhouse Gases Program, established under Section 1605(b) of the Energy Policy Act of 1992, records the results of voluntary measures to reduce greenhouse gas emissions. For the 2005 reporting year, 221 US companies and other organizations reported to the Energy Information Administration (EIA) that they had undertaken 2,379 projects to reduce or sequester greenhouse gases in 2005. The 1605(b) voluntary registry was the first of its kind in the US, but the results of the registry have not been well received by environmental and other organizations interested in fostering real greenhouse-gas emissions reductions. The fundamental problem with the registry in terms of accounting for real reductions is that the program allows firms to report on successful emissions reduction projects, while remaining silent on whether their overall emissions levels have increased or decreased. A recent study conducted at the University of Michigan found that for electricity-generating companies the program has no statistically significant effect on a firm's carbon intensity (i.e., its carbon emissions per unit of electricity generated).[23]

Successful implementation of a registry that could provide the basis for granting emissions credit in a regulatory program or providing the basis for credit sales would require careful consideration of how to ensure registered credits are real, verifiable, and additional (e.g., not "anyway" tonnes, i.e., credits given for doing what the company would do anyway). California is currently working on developing a registry that might serve as a template for a national program.

Allocation of Government Incentives

The Energy Policy Act of 2005 contains a number of incentives, including investment tax credits and federal loan guarantees. Allocation of effective incentives that become over-subscribed is a difficult policy issue. Investment tax credits for IGCC have been over-subscribed by as much as three times (for bituminous fuel), and for industrial gasification by over seven times. The loan guarantees for gasification contained in the Energy Policy Act of 2005 have not yet been fully implemented by the responsible federal agency, seven years after the passage of the legislation.

Both issues are critical for investors in these multi-billion dollar projects. Implementation of the 2005 Energy Policy Act provisions should be a very high priority. The three applications for gasification loan guarantees selected on

October 8, 2007, should be expeditiously processed and similar loan guarantees should be implemented for additional projects.

Allowing Flexibility During Plant Operations

Unlike the decision to build a plant, which is based on achieving a certain return from the funds invested, under a tax or a cap and trade scheme the decision to operate a plant is based on whether the marginal operating costs can be covered by the revenue generated. Recovering the capital costs is not included in that decision, since at that time they have become "sunk costs." So an expensive CCS-capable facility that was constructed based on an expected regulatory future may end up operating despite being unable to meet its loan payment schedule.

When compared to the variability of the price of electricity, which can fluctuate dramatically over the course of a day (from peak to off-peak periods), over the course of a week (from weekdays to weekends), and over the course of a year (from summer to winter), the cost of emitting CO_2 is likely to be comparatively stable (set by either a tax or market). This could create times of the year when the return from not using the CCS equipment (and incurring the significant associated energy costs) could more than cover the extra costs that venting CO_2 to the atmosphere and paying for the emissions would incur. This, of course, depends on:

- what the plant is allowed to do (e.g., can it operate while venting CO_2?);
- the ability of the plant to curtail part of its operations (e.g., turn off the compressors or pumps); and
- the ability to regain the "saved" energy (either in increased electricity output or in reduced heat rate).

Given the uncertainty in the electricity and carbon markets, having the "option" to operate the plant in one of four ways (i.e., off, full-CCS, partial capture but no sequestration, or as a plant with no CCS) could have real value to the operator and investor. Depending on its magnitude, this option value could be sufficient to help encourage "fence sitters" to invest.

A fundamental insight from financial option theory is that without uncertainty an option has no value and, depending on the problem, increasing uncertainty increases the value of the option. If the long-term future carbon price was known with certainty, then the need for an option that allowed a CCS facility to operate in different configurations would be very low. High carbon prices would favor a plant that could capture and sequester a large percent of its carbon emissions, while low prices would favor a plant that vented and "paid the penalty." To demonstrate this option value, a simple case study was developed for a new 500 MW pulverized coal plant. Figure 9.1 shows the performance (annual operating profit) for these two designs for different carbon costs (in dollars per tonne CO_2).

Determining the value of having the option to operate the plant in different ways requires engineering and financial modeling skills. As compared to a

FIGURE 9.1 *Operating profit for two PC 500 MW plants: (1) with no CCS equipment venting 100% and paying the CO_2 price per tonne, and (2) with 90% CO_2 capture and sequestration.*

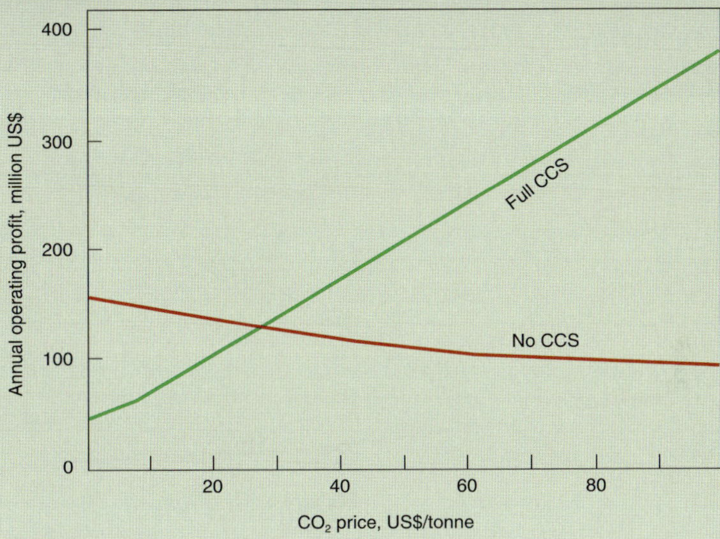

design constructed with a goal of capturing as much CO_2 as possible, a design that provides flexibility in how a plant can operate could look very different. It might be financially beneficial to "give up" some capture effectiveness in order to gain operational flexibility. For example, while an IGCC plant with full CCS might provide the most cost-effective way to capture large portions of the carbon emission, a PC plant with post-combustion capture might provide more options for regaining energy if venting.

Figure 9.2 shows the profit profile for a plant that could either fully sequester or vent and capture 50% of the energy saved. If the uncertainty in the carbon price follows the distribution shown, the option value exceeds by over 5–10% the profits earned from either a full-vent or full-CCS plant depending on the distribution assumptions.

Numerous assumptions, both engineering and financial, lie behind this preliminary case study. In addition, the operating characteristics of plants when cycling between venting and CCS modes need to be explored. The more rapidly the plant can react to changes in electricity and carbon markets, the more profitable the "option" to operate in different configurations will be. However, because of engineering limitations, it may be impractical if not impossible to cycle on an hourly or even daily basis. Also critical to this analysis is how the carbon costs are "passed through" to the electricity prices. A price of $50 per tonne for emitted CO_2 will certainly affect the cost of producing a MWh from a coal plant and to

FIGURE 9.2 *Operating profit for two PC 500 MW plants: (1) with no CCS equipment venting 100% and paying the CO_2 price per tonne, (2) with 90% CO_2 capture and sequestration, and (3) option of operating in a full-vent or full CCS configuration. Also shown is the distribution of expected future CO_2 prices.*

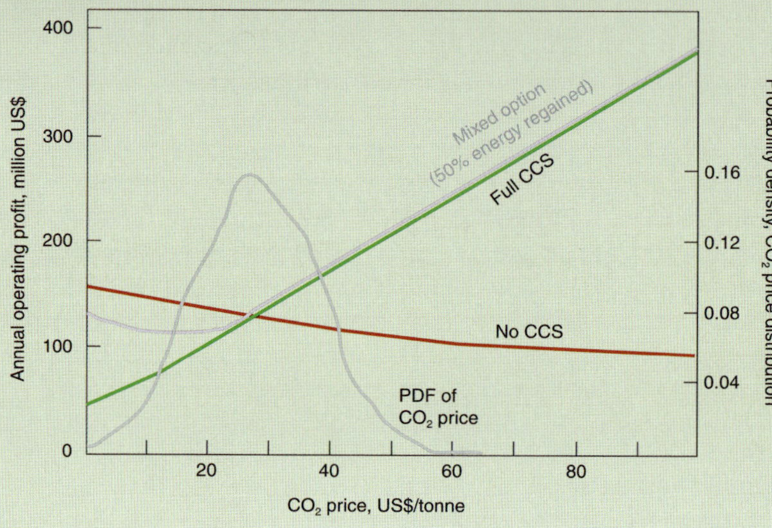

a lesser degree from a gas turbine plant. It is unclear how this would shift the off-peak and peak prices. If most, if not all, of these additional costs pass through, then the penalty for venting is minimized and the benefits for CCS are small. This pass-through percent would vary across the country because of differences in generating technology mixes. Any pass-through would also affect demand for electricity.

Regulations could be written that block such a flexible operating scenario from occurring. Regulators could force the CCS-equipped plants from venting by requiring a certain captured tonnage or percent per year in order to gain tax incentives or other advantages. Such a limitation would probably promote more rapid reduction in CO_2 emissions but might act to slow private capital investment in CCS.

Looking out further, regulations and carbon taxes and markets will have medium- to long-term effects on commodity markets (e.g., potentially decreasing coal prices and increasing natural gas prices). This in turn will affect future new plant construction. The recent drop in natural gas prices caused by the economic downturn has caused some gasification projects to be reconsidered. This could, in turn, lead to an increased demand for natural gas and subsequent increase in its price.

9.2 FOSTERING TECHNOLOGY INNOVATION

Through its influence on markets for environmental technologies (such as CO_2 capture and storage systems), government actions also are a critical element of the technological innovation process. Studies of past measures to reduce sulfur dioxide and nitrogen oxide emissions from US power plants have shown that regulatory policies influence both the magnitude and the direction of efforts to develop new or improved capture technologies. For example, Figure 9.3 shows the century-scale trends in US patenting activity for SO_2 reduction technologies[24] and

FIGURE 9.3 *History of US patent activity for SO_2 control technology (above) and for NO_x control technology (below). In both cases, patent activity shows a dramatic increase in response to or anticipation of regulatory activity. (Figure redrawn and updated from Taylor et al.,* Technological Forecasting and Social Change *and Yeh et al.,* Journal of the Air & Waste Management Association.*)*

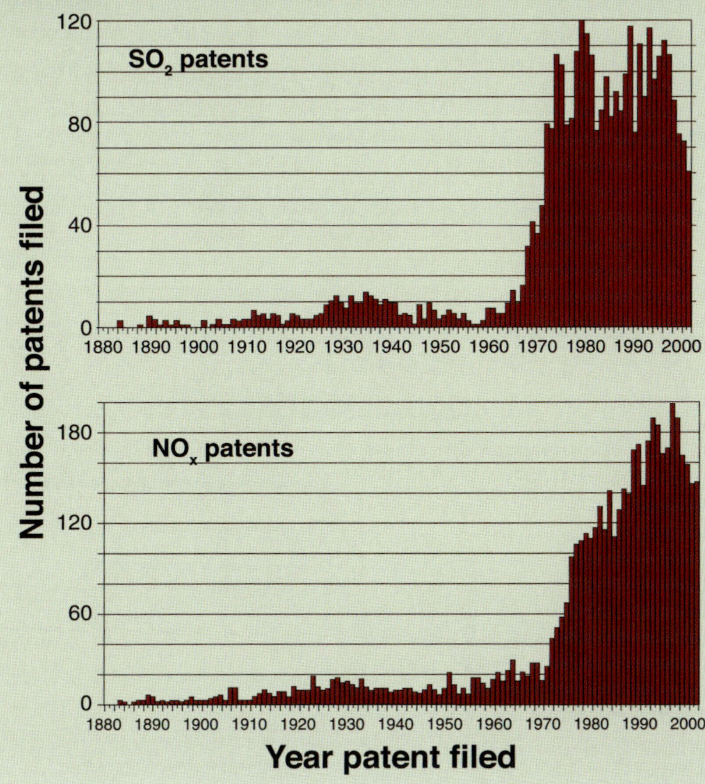

post-combustion NO_x capture systems,[25] respectively. In both cases, the number of new patents filed, a measure of "inventive activity," increased dramatically when new environmental regulations that required or incentivized the use of these technologies came into force. (In the case of NO_x control, such regulations for coal plants came first in Japan and Germany; US regulations lagged by more than a decade.)

While most of us are used to thinking of learning curves as starting with high costs and then monotonically decreasing with experience and volume, in the US in the case of both SO_2 and NO_x control, costs first went up before finally starting down,[26] as problems were discovered with the first several plant designs (Figure 9.4). One can reasonably anticipate a similar experience with commercial-scale CO_2 capture technology.

FIGURE 9.4 *While most of us expect learning or experience curves to start high and then slope monotonically downward (A), in the case of both SO_2 (B, above) and NO_x control (B, below), costs rose before finally starting down. It is reasonable to expect a similar experience with CO_2 capture technology. This means that it is important to start building commercial-scale capture systems today, so that they will be ready, reliable, and cost-effective when they are needed. (Figure B modified from Rubin et al.,* International Journal of Greenhouse Gas Control.*)*

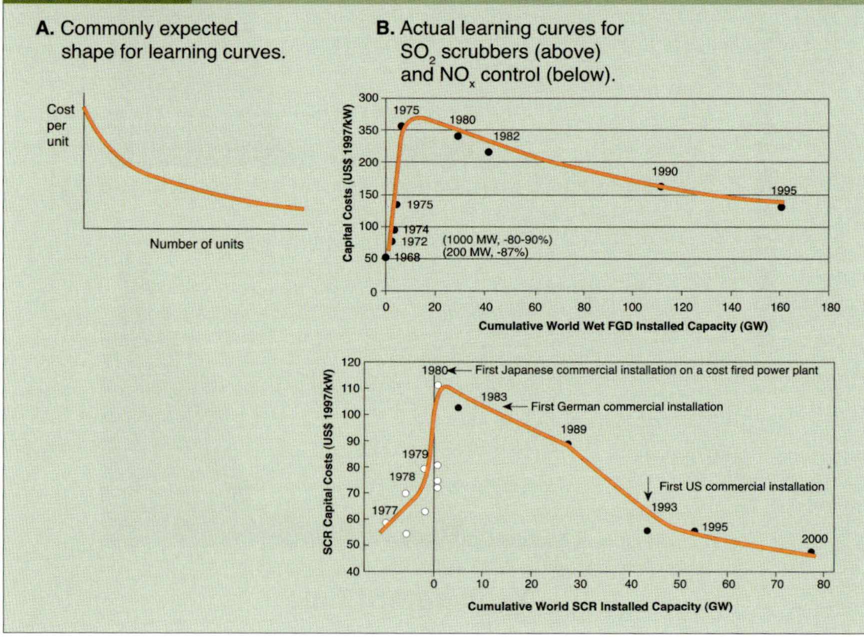

Thus, while strong and sustained support for research and development is essential to the development of new, improved, and lower-cost processes for carbon capture and sequestration, there is no substitute for beginning to build commercial-scale plants. That is not likely to occur without a policy or regulatory framework that fosters innovation through the creation of markets for improved, lower-cost CCS technologies.

9.3 SUMMARY

This chapter has outlined some of the current concerns and constraints facing widespread commercialization of carbon capture and sequestration. To help make CCS a commercial reality, measures such as government loan guarantees, tax credits, direct financial subsidies, and R&D funding can play an important role. Such measures help to reduce financial and other risks, and create markets for CCS by providing incentives for technology development, deployment, and diffusion. At the same time, government regulatory policies also play a central role in creating future markets for CCS technology. Careful design of a regulatory framework also is essential to avoid unintended consequences that could lead to undesirable or suboptimal outcomes while trying to incentivize CO_2 sequestration from power plants. The earlier chapters of this book have presented an approach that we believe can achieve that objective.

10

CONCLUSIONS AND RECOMMENDATIONS

If the US is going to reduce emissions of carbon dioxide by an order of magnitude over the next fifty years, large-scale deployment of carbon capture with deep geologic sequestration (CCS) will be a necessary part of a portfolio of strategies. Two things must happen before large-scale deployment of CCS can occur: 1) the technology must be further developed, and made cost-effective at commercial scale (i.e., in plants with capacities of many hundreds of MW); and 2) dramatic improvements must be made in what is now an incomplete and inadequate regulatory system. This book has laid out our vision of what such a regulatory framework should be.

CREATE A FEDERAL OPT-IN PROGRAM FOR CO_2 PIPELINES

Because geologic formations that are suitable for sequestration are sometimes located far from the best sites for power plants, a network of carbon dioxide pipelines, perhaps as large or larger than the present half-million kilometer natural gas network, will be required if large-scale use of CCS is to become a reality. Today there are over 5,600 km of carbon dioxide pipelines in the US. Federal safety rules apply to all pipelines. However, in contrast to natural gas, all of the present CO_2 pipelines have been sited, built, and operated under state law. Future pipeline developers should remain free to use these existing state-centric frameworks to build additional pipelines. However, such arrangements are not likely to be adequate as the scale of deployment grows. In Chapter 3, we outlined a new statutory arrangement that would permit the Federal Energy Regulatory Commission (FERC) to supervise an "opt-in" program for federal siting and economic regulation of carbon dioxide pipelines built specifically for CCS. Under the "opt-in" program that we have proposed, operators would be able to apply to FERC for federal siting authority, which, if granted, would permit the use of federal eminent domain authority to site the pipelines. The FERC permit would obligate the pipelines to operate as common carriers with non-discriminatory access, but without FERC or other regulation of rate structures. Regardless of which scheme developers adopt, we have also recommended a streamlined permitting process for pipelines across federal lands.

ESTABLISH AN INTEGRATED FEDERAL APPROACH TO GEOLOGIC SEQUESTRATION

In December 2010, the US Environmental Protection Agency (EPA) promulgated a new rule for CO_2 sequestration wells, by creating the Class VI well category under the existing Underground Injection Control (UIC) regulatory program. This new class applies only to wells that inject CO_2 for geologic sequestration. Wells injecting CO_2 for enhanced oil recovery (EOR) will continue to be treated as Class II wells. Compared to past UIC regulations, the new Class VI rules have more adaptive elements and incorporate some of the advances in modeling and monitoring that have emerged in the thirty years since the first UIC rules were promulgated. However, because none of the regulatory requirements imposed on other well classes have been similarly updated, the Class VI rules can be read as disproportionately stringent.

Although the EPA now has the authority under the UIC program to permit and regulate geologic sequestration of CO_2, the agency currently has no authority to resolve a variety of other issues that must be addressed if CCS is to become safe, environmentally sound, affordable, compatible with evolving national and international CO_2 control regimes, and socially equitable.

We believe that a comprehensive national legal and regulatory framework must be created if geologic sequestration is to be widely developed as part of a portfolio of strategies to reduce the emissions of greenhouse gases. New federal regulation should provide structure and certainty for CCS project developers, investors, and the public, while ensuring that injection operations are safe and permanent. This new framework should assure access to pore space, establish arrangements for the management of long-term stewardship and liability for sequestration sites once they have been closed, and set up greenhouse gas (GHG) accounting protocols. The legislation should declare that sequestering carbon dioxide in appropriate geologic formations, in order to mitigate the detrimental effects of climate change, is in the public interest. It also should amend the Safe Drinking Water Act (SDWA) to authorize regulators to balance protection of underground sources of drinking water (USDW) with other compelling national interests, such as mitigating climate change.

We elaborated these arguments in Chapters 4 through 7. The Appendix contains suggested legislative language to implement the needed changes.

LEARN AND ADAPT WITH EXPERIENCE

Because CCS has not yet been adopted on a large scale, and because the effects of injecting large volumes of fluid into the deep subsurface will always be somewhat uncertain, legislators and regulators will not be able to fully anticipate all of the contingencies that may arise, or the lessons that will come as more and more field experience accumulates.

Two strategies hold promise to keep regulation effective and efficient as more is learned about geologic sequestration. *Performance-based regulation* specifies the results that should be achieved, rather than the means by which they are to be achieved. It is an alternate approach to design- or technology-based regulation that specifies particular behaviors, technologies, procedures, or processes. *Adaptive regulation* entails a deliberate process of periodic review and revision that uses experience and new knowledge to clarify and improve the rules. Adaptive regulation should not be automatically retroactive; it should not change the regulatory requirements for existing projects unless operators and regulators both agree. It can, however, adapt the rules so that new permits are granted based on the best current scientific and technical knowledge.

After describing these two regulatory approaches in greater detail, and providing examples of both, in Chapter 5 we recommended that Congress enact new legislation on geologic sequestration that encourages the EPA to promote performance standards for CCS wherever that is feasible, and mandate periodic review and revision of all geologic sequestration regulations. To achieve the latter, we have recommended that the new legislation establish and fund an independent Technical Advisory Committee (TAC) under the auspices of the National Research Council (NRC). This committee would be charged with periodically reviewing accumulated GS operation experience from around the world, and using this accumulated knowledge to advise regulators (the EPA, a proposed Federal Geologic Sequestration Board in the US Department of Energy, and relevant state agencies) on ways to improve the regulatory framework for GS. Such a review should occur at least once every seven years.

ESTABLISH A FEDERAL FRAMEWORK FOR ACCESS TO PORE SPACE

When carbon dioxide is injected underground it fills the "pore space" between grains in rock, displacing very briny water. When other kinds of waste are injected in the US today under the UIC program, nothing is done to obtain permission to use this pore space from property owners. However, given the very large capital investments involved in CCS, and the possibility that some injection projects could acquire high public visibility, we believe that the issues of legal access to and use of pore space must be clearly resolved if CCS is to be implemented on a large scale. We are primarily concerned about the right to inject CO_2 into suitable deep geologic reservoirs, as opposed to the more complicated question of who "owns" the pore space in such reservoirs.

In the United States, state law has historically defined property rights. Today, from state to state, a patchwork of legal arrangements governs access to and use of pore space. In Chapter 6, we described and assessed a number of alternative models that might be used to address this issue. These include the regimes now in use for natural gas storage and for underground waste injection under UIC, and a model based on current practice in groundwater aquifer storage and recharge. We also described and evaluated the various ways in which the issue of access to pore space might be resolved through legislation and the courts.

While our concern is with access and use, Wyoming, Montana, and North Dakota have passed legislation explicitly defining pore-space ownership. The Takings Clause of the Fifth Amendment to the US Constitution instructs that private property shall not "be taken for public use, without just compensation." However, there is clear legal precedent that if a taking imposes a *de minimis* burden on a property owner, then compensation can be nominal.

After reviewing a wide range of options, we have concluded that federal legislation should codify a formal process for managing the access and use of pore space for geologic CO_2 sequestration. Under this process, a project developer could apply to the appropriate UIC agency (either federal, or a state under delegated authority) in order to acquire both an injection permit and a permit to use the pore space for sequestration. Under the arrangement we detailed in Chapter 6, once a project developer had applied for a pore-space use permit, there would be suitable opportunity for notice and comment to allow other operators to intervene and file a competing application and to resolve issues of potentially conflicting uses. Once approved, the use permit would provide an exclusive right to access and use a specified region of pore space for geologic CO_2 sequestration, and would provide protection from trespass liability within that region. In Chapter 6, we outlined a very similar process for granting injection and pore-space permits on federal land.

Nothing in the approach we have proposed would prohibit an operator from acquiring pore space rights through private negotiations with surface landowners. Indeed, if a project developer acquired consent to use pore space from all property owners overlying the subsurface project boundary defined in the CO_2 injection permit, a pore-space use permit would not be necessary.

ESTABLISH A FEDERAL STRATEGY FOR LONG-TERM STEWARDSHIP AND LIABILITY

We believe that the liability issues that may arise during the pre-operational, operational, and immediate post-closure phases of a project to sequester CO_2 will not be dramatically different from those associated with many other large industrial projects, and can be readily handled through conventional insurance/mechanisms. In contrast, we believe that the issues associated with long-term stewardship of an injection site that has been closed, including oversight, possible remediation, and liability, will require a special regulatory framework. To manage these issues, in Chapter 7 we have proposed the creation of a Federal Geologic Sequestration Board (FGSB), administratively housed in the US Department of Energy. This Board would have responsibility for approving sites to be accepted into long-term stewardship, and for overseeing any remedial actions that may be required. Funding for this Board and its activities would come from a Carbon Sequestration Trust Fund into which all operating projects would pay, at rates based on several criteria, including the number of tonnes of carbon dioxide they sequester.

ESTABLISH A CO_2 ACCOUNTING FRAMEWORK FOR ALL PHASES OF CCS ACTIVITIES

Under almost any program to limit the emission of greenhouse gases, accounting protocols will be needed to document the amount of CO_2 that is sequestered as well as the CO_2 emissions that have been avoided. Because there are limits to the precision with which the volume of injected CO_2 can be measured, and much greater limits on how accurately the amount of CO_2 that remains in sequestration can be determined, developing approaches to accounting that are technically feasible, yet adequate for regulatory and other necessary oversight, poses significant challenges.

We have discussed many of these challenges in Chapter 8. There we concluded that only CO_2 injected at a site with an approved monitoring plan should be considered geologically sequestered under any federal program for greenhouse gas reduction. Because routine subsurface monitoring required by injection permits will include methods that can detect loss of containment, we believe that routine surface monitoring, to prove that leakage is not occurring, should not be required. However, if routine monitoring of the subsurface does detect loss of containment, then GS site operators should be required to undertake a targeted monitoring program to measure surface leakage emissions from the site. Such a program should only be required if subsurface monitoring indicates CO_2 has migrated through the confining formation and surface monitoring detects leakage.

If subsurface monitoring detects loss of containment, operators will need to know the performance standard they must meet to detect, locate, and measure surface leakage. Thus, we have argued that regulators should set technically plausible detection limits for surface leakage. Regulators should also develop a backstop protocol to compensate for surface leakage in cases where monitoring and measurement are technically infeasible or inaccurate. An integrated CO_2 inventory tracking system should be established, and parties responsible for any CO_2 lost or unaccounted for in handling should be required to take appropriate compensatory action.

In Chapter 8, we recommended that GS projects, along with CO_2 capture facilities and pipelines, be classified as emissions sources under any future cap and trade program. This approach would cleanly define the boundaries between the three stages of the CCS process (which are likely to be run by different commercial entities), and clearly accounts for GHG emissions from energy used to capture, compress, transport, and inject CO_2. As noted above, in order to cover the cost of potential surface leakage during long-term stewardship, GS site operators should contribute to a Carbon Sequestration Trust Fund.

STOP PROCRASTINATING AND MAKE CCS A REALITY

Without a climate or energy program that is committed to significantly reducing GHG emissions, there will be no commercial-scale CCS, because it will always be cheaper to vent CO_2 to the atmosphere than to capture and sequester it. At the same time, it could take a decade or more of experience with large-scale pilot projects before CCS is perfected and becomes economically viable. In Chapter 9, we discussed and evaluated a variety of strategies that might be employed to accelerate the development of CCS so that the technology will be available when the US is ready to undertake a serious program of emission reduction. Strategies considered included direct government project funding, government subsidies, a variety of different kinds of tax credits, government payments per metric ton of CO_2 sequestered made to early operators, arrangements by state Public Utility Commissions to facilitate early adoption, the creation of a CCS trust fund, and several others.

The technical, legal, and policy analysis that we have completed for this book has focused on developing a framework for the regulation of CCS. Thus, we have not made firm recommendations about which policies should be adopted to promote the rapid development and early adoption of CCS technology. But we firmly believe that, in addition to the regulatory framework we have articulated, the adoption of several such policies is urgently needed.

For too long, nations around the world, including the US, have talked about CCS without taking action. In the meantime, the atmospheric concentration of carbon dioxide and other greenhouse gases has continued to grow, and changes in climate have become ever more apparent. The time has come to stop talking. The US and the world need to move quickly to develop and deploy CCS in a way that is safe, environmentally sound, affordable, compatible with evolving national and international CO_2 control regimes, and socially equitable. We hope that the ideas developed in this book will help speed up that process.

APPENDIX

A BILL

To establish a comprehensive system for the safe and effective transport and geologic sequestration of carbon dioxide.

Be it enacted by the Senate and House of Representatives of the United States of America in Congress assembled,

SECTION 1. SHORT TITLE.

This Act may be cited as the "Carbon Capture and Sequestration Regulatory Act of 2012".

SEC. 2. TABLE OF CONTENTS.

The table of contents for this Act is as follows :

SEC. 3. FINDINGS.

The Congress finds that facilitating the sequestration of carbon dioxide in appropriate geologic formations in order to mitigate the detrimental effects of climate change is in the public interest. The Congress further finds that a nationally-coordinated flexible and adaptive regulatory approach is needed to assure that such sequestration is undertaken in a manner that is safe, environmentally sound, affordable, socially equitable, and compatible with evolving international CO_2 control regimes.

SEC. 4. DEFINITIONS.

For the purposes of this Act:

(a) GEOLOGIC SEQUESTRATION FACILITY.—The term "geologic sequestration facility" or "GS facility" means a facility that receives and permanently stores or sequesters carbon dioxide in a geologic formation. Such term includes an enhanced hydrocarbon recovery operation only if a CO_2 injection permit has been issued for such project.

(b) LONG-TERM STEWARDSHIP.—The term "long-term stewardship" means the monitoring, measurement, verification, and remediation and related activities associated with a GS project after issuance of a certificate of closure.

(c) CARBON DIOXIDE.—The term "carbon dioxide" or "CO_2" means carbon dioxide that has been captured from the air or from an emission source, including incidental substances derived from the source materials and the capture process, and any substances added to the stream to enable or improve the injection process.

(d) ADMINISTRATOR.—The term "Administrator" means the Administrator of the Environmental Protection Agency.

(e) BOARD.—The term "Board" refers to the Federal Geologic Sequestration Board that is established under title IV of this act.

(f) CCS.—The term "CCS" refers to carbon capture and sequestration, which is the process of capture, transport, and geologic sequestration of CO_2 for the purpose of permanently preventing its release to the atmosphere.

(g) ENHANCED HYDROCARBON RECOVERY.—The term "enhanced hydrocarbon recovery" or "EHR" means injection of CO_2 for the purpose of increasing production of oil or natural gas.

(h) GEOLOGIC FORMATION.—The term "geologic formation" refers to a body of rock in the subsurface that is distinct from the surrounding rock and is capable of permanently retaining injected carbon dioxide.

(i) PORE SPACE.—The term "pore space" means the voids in a geologic formation that can contain a fluid.

(j) REMEDIATION.—The term "remediation" means any activity undertaken at any stage during the operation, closure or long-term stewardship of a sequestration site that is intended to correct a health, safety, or environmental problem with the performance of a site such as undesired movement of the injected fluid, or possible release of carbon dioxide to the atmosphere.

(k) SITE OPERATOR.—The term "site operator" refers to the entity responsible for the operation of the geological sequestration facility.

(l) CO_2 PIPELINE.—The term "CO_2 pipeline" means any pipeline constructed for, or converted for, use in the transport of carbon dioxide for purposes of geologic sequestration.

SEC. 5. SEVERABILITY OF PROVISIONS.

If any provision of this act, or the application of any provision of this act to any person or circumstance, is held invalid, the application of such provision to other persons or circumstances and the remainder of this act shall not be affected thereby.

TITLE I—CARBON DIOXIDE PIPELINES

SEC. 101. SITING AND CONSTRUCTION OF CO_2 PIPELINES.

(a) Definitions.—For purposes of this title:

(1) CO_2 Pipeline project.—The term "CO_2 pipeline project" means a project to construct a CO_2 pipeline after the date of enactment of this section.

(2) Certificated CO_2 pipeline.—The term "certificated CO_2 pipeline" project means a CO_2 pipeline project for which a certificate has been issued under subsection (b)(3).

(3) Commission.—The term "Commission" means the Federal Energy Regulatory Commission.

(b) Elective Procedure for Siting and Construction of CO_2 Pipeline Projects.—

(1) Application.—Any person may apply to the Commission for a certificate of public convenience and necessity that authorizes such person to:

(A) construct, acquire or operate a CO_2 pipeline project, or

(B) modify a certificated CO_2 pipeline project.

(2) Application procedure.—

(A) Application for a certificate shall be made in writing to the Commission, and shall be in such form, contain such information, and notice thereof shall be served upon such interested parties (including State Commissions), in such manner as the Commission shall, by rule, require. The Commission shall set the matter for hearing and shall give notice of the hearing to interested persons; and the application shall be issued or denied in accordance with the procedure provided in paragraph (3).

(B) Notwithstanding subparagraph (A), the Commission may issue a temporary certificate in cases of emergency, to assure maintenance of adequate service or to serve particular customers, without notice or hearing, pending the determination on an application for a certificate, and may by regulation exempt from the requirements of this subsection acts or operations for which the issuance of a certificate will not be required in the public interest.

(3) GRANT OF CERTIFICATE.—Except in the cases governed by paragraph (2)(B), a certificate shall be issued to a qualified applicant therefore, authorizing the whole or any part of the operation, construction, acquisition, or modification covered by the application, if it is found that the applicant is able and willing properly to do the acts and to perform the service proposed and to conform to the provisions of this title and rules of the Commission hereunder, and that the proposed operation, construction, acquisition, or modification, to the extent authorized by the certificate, is or will be required by the present or future public convenience and necessity; otherwise such application shall be denied. The Commission shall have the power to attach to the issuance of the certificate, and to the exercise of the rights granted thereunder, such reasonable terms and conditions as the public convenience and necessity may require.

(4) OPEN ACCESS AND RATE CONDITIONS.—In granting a certificate under paragraph (3), the Commission shall include conditions under paragraph (3) that are designed to ensure—

(A) That the capacity of the pipeline is sufficient to accommodate all potential shippers that execute legally binding undertakings (including adequate collateral or evidence of creditworthiness) to pay a contractually established rate for requested capacity for a term necessary to support financing of the pipeline.

(B) That any unused capacity of the pipeline is made available on a non-discriminatory basis.

(C) In cases where the pipeline and the shipper cannot agree on rates and terms and conditions of service, and the rates and terms and conditions of service are not unreasonable, the Commission may modify conditions of a certificate if it determines on the record after hearing that the public convenience and necessity so require.

(5) RIGHT OF EMINENT DOMAIN.—When any holder of a certificate issued under paragraph (3) cannot acquire by contract, or is unable to agree with the owner of property to the compensation to be paid for, the necessary right-of-way to construct, operate, and maintain the project to which the certificate relates, and the necessary land or other property necessary tothe proper operation of such project, it may acquire the same by the exercise of the right of eminent domain in the district court of the United States for the district in which such property may be located, or in the State courts. The practice and procedure in any action or proceeding for that purpose in the district court of the United States shall conform as nearly as may be with the practice and procedure in similar action or proceeding in the courts of the State where the property is situated.

(c) FEDERAL AGENCY COORDINATION.—

(1) RIGHTS OF WAY ON FEDERAL LANDS.—In the case of any CO_2 pipeline project for which an application is filed under subsection (b)(1) or granted under subsection (b)(3), the Commission shall exercise authority over such project comparable to that which it exercises under section 15 of the Natural Gas Act with respect to natural gas pipelines.

(2) NEPA REVIEW.—The Commission shall be lead Federal agency for purposes of Federal environmental review and, in consultation with affected agencies, shall prepare a single envi-

ronmental review document that would be used as the basis for all decisions under Federal law related to the proposed project.

(3) ADMINISTRATIVE POWERS OF THE COMMISSION.—The Commission shall have the same authority under sections 8 through 25 of the Natural Gas Act with respect to certificated CO_2 pipelines and owners and operators of such pipelines as it has with respect to natural gas pipelines and natural gas companies.

SEC. 102. SAVINGS PROVISIONS.

(a) STATE REGULATION.—Nothing in this title shall modify or limit State siting and economic regulation of any CO_2 pipeline that is not a certificated CO_2 pipeline.

(b) PIPELINE SAFETY.—Federal pipeline safety regulation under chapter 601 of title 49, United States Code, applies according to its terms to CO_2 pipelines, whether or not they are certificated CO_2 pipelines.

TITLE II—ADAPTIVE AND PERFORMANCE-BASED APPROACH

SEC. 201. NEED FOR AN ADAPTIVE AND PERFORMANCE-BASED APPROACH TO REGULATING THE GEOLOGIC SEQUESTRATION OF CARBON DIOXIDE.

(a) THE CONGRESS FINDS AND DECLARES:

(b) IN GENERAL.—Because at present there is limited experience with large-scale geologic sequestration of carbon dioxide, regulations developed to govern GS site permitting operation and issuance of a certificate of closure, including criteria for a transition to long-term stewardship, should emphasize the promulgation of adaptive and performance-based rules and standards to encourage flexibility and learning, and mandate a systematic process to regularly review and improve the regulation of GS.

(c) DEVELOPMENT OF PERFORMANCE-BASED RULES AND STANDARDS.—The objective of rules and standards developed by the Environmental Protection Agency for permitting and monitoring the operation of GS Projects shall be to assure that these activities are conducted in a manner that provides reasonable protection to health, safety and the environment. Similarly, the objective of the standards or criteria developed by the Federal Geologic Sequestration Board for accepting GS sites into long-term stewardship and managing those sites once they have entered long-term stewardship shall be to assure that these standards, criteria and management activities continue to provide reasonable protection to health, safety and the environment. Consistent with those objectives, both the Environmental Protection Agency and the Federal Geologic Sequestration Board shall, to the extent practicable, promulgate rules and formulate standards that are performance-based such that compliance strategies may evolve with increased knowledge, while the required level of regulatory performance remains constant.

SEC. 202. CREATION OF A CCS TECHNICAL ADVISORY COMMITTEE OF THE NATIONAL RESEARCH COUNCIL.

(a) FORMATION OF THE TECHNICAL ADVISORY COMMITTEE.—Within three months of the adoption of this Act, the EPA shall contract with the National Research Council (NRC) to establish an independent expert CCS Technical Advisory Committee. Membership of this committee shall consist of at least nine experts, chosen by the NRC for their technical expertise in accordance with standard NRC procedures.

(b) INITIAL REVIEW.—Within one year of its establishment, the CCS Technical Advisory Committee of the NRC shall complete a review of the available accumulated data and experience from operational carbon dioxide sequestration projects in the U.S. and elsewhere, and publish a summary of its findings together with recommendations, based on those findings, as to:

(1) how regulations and standards for permitting and operating a carbon dioxide geologic sequestration facility should best be shaped to provide reasonable protection to health, safety and the environment;

(2) the feasibility of adaptive strategies and performance-based standards that improve the effectiveness of GS operations with no loss of protection to health, safety and the environment; and,

(3) research needed to provide the foundation for improving standards in the future.

(c) PERIODIC REVIEW.—No later than seven years after the NRC CCS Technical Advisory Committee's initial review, and at least every seventh year thereafter, the Administrator and the Chairperson of the Federal Geologic Sequestration Board shall jointly request that the Committee be reconvened to:

(1) update and evaluate the cumulative experience from all geologic sequestration facilities operating in the U.S. or elsewhere;

(2) identify research to provide the foundation for improving the formulation of standards in the future; and,

(3) publish a summary of its findings and make recommendations regarding the following:

(A) modifications to performance-based elements of the regulations, such as clearer articulation of objectives, specification of performance standards, guidance on assessment methodologies to insure that performance standards are met, and requirements for overall accountability;

(B) modifications to the prescriptive elements of the regulations to make the requirements reflect the best current science, or to advance them toward being more performance-based;

(C) revise how the regulatory authority for GS is balanced between EPA programs and Federal and State regulators, concentrating on streamlining oversight, reducing duplication and eliminating regulatory gaps;

(D) propose new legislation, or amendments to existing laws that could improve the environmental performance, economic efficiency, or GHG reduction capacity of GS; and,

(E) any other necessary changes based on scientific, technologic and economic developments.

(d) FUNDING MECHANISM.—Funding for the first review by the CCS Technical Advisory Committee of the NRC shall be provided from the operating budget of the Environmental Protection Agency in an amount not to exceed $800,000. Funding for subsequent reviews by the CCS Technical Advisory Committee shall be provided from the carbon sequestration trust fund established under section 403(d) in an amount jointly determined to be appropriate by the Administrator and the Chairperson of the Federal Geologic Sequestration Board.

SEC. 203. REQUIREMENTS FOR PERIODIC REVIEW AND REVISIONS.

(a) PERIODIC REVIEW AND REVISION OF REGULATIONS FOR GS PROJECTS.—

(1) Within eight years from the date of the NRC CCS Technical Advisory Committee's initial review conducted under section 202(b), and at least every eighth year thereafter, the Environmental Protection Agency shall review, and if necessary, revise the details of the regulations and standards it has promulgated for GS projects.

(2) Three months before undertaking the review noted in subpart (1), the Administrator of the Environmental Protection Agency shall publish notice in the Federal Register and provide an opportunity for public comments to be filed.

(3) The Environmental Protection Agency's review shall be informed by the summary of accumulated experience developed by the CCS Technical Advisory Committee of the National Research Council; by any recommendations that the Committee may make; and by any public comments received by the Administrator as a result of the call for public comments. In accordance with the requirements of the Administrative Procedures Act, a description of any resulting changes shall be published in the Federal Register and adopted only after public comment and review. As part of this review, and consistent with the objective of assuring reasonable protection to health, safety and the environment, the Agency shall consid-

er whether any procedure-based regulation it previously promulgated should be replaced by a performance-based regulation.

(b) Periodic Review And Revision Of Procedures Governing Long-Term Stewardship.—

(1) Within eight years from the date of the NRC CCS Technical Advisory Committee's initial review conducted under section 202(b), and at least every eighth year thereafter, the Federal Geologic Sequestration Board shall review and, if necessary, revise in light of the recommendations and accumulated experience reviewed by the CCS Technical Advisory Committee of the National Research Council, the details of its rules for accepting GS projects into long-term stewardship, and its procedures for monitoring and otherwise managing projects held in long-term stewardship.

(2) Three months before undertaking the review noted in subpart (1), the Board shall publish notice in the Federal Register and provide an opportunity for public comments to be filed.

(3) The Board's review shall be informed by the summary of accumulated experience developed by the CCS Technical Advisory Committee of the National Research Council; by any recommendations that the Committee should make as part of its report; and by any public comments received by the Administrator as a result of the call for public comments. In accordance with the requirements of section 553 of title 5, United States Code, a description of any resulting changes shall be published in the Federal Register and adopted only after public comment and review. As part of this review, and consistent with the objective of assuring reasonable protection to health, safety and the environment, the Board shall consider whether any procedure-based regulation it previously adopted should be replaced by a performance-based regulation.

SEC. 204. JUDICIAL REVIEW.

Nothing in this title shall affect any otherwise available judicial review of agency action. This title is intended only to improve the internal management of the regulation of GS projects, and does not create any right or benefit, substantive or procedural, enforceable at law or equity by a party against the United States, its agencies or instrumentalities, its officers or employees, or any other person.

TITLE III—PERMITS AND PORE SPACE ACCESS FOR GEOLOGIC SEQUESTRATION PROJECTS

SUBTITLE A—GENERAL

SEC. 301. DEFINITIONS.

For the purposes of this title:

(a) CO_2 INJECTION PERMIT.—The term "CO_2 Injection Permit" refers to a UIC permit issued by the UIC CO_2 Regulator under subtitle B for the permanent geologic sequestration of carbon dioxide.

(b) INTERSTATE GEOLOGIC SEQUESTRATION PROJECT.—The term "interstate geologic sequestration project" means a geologic sequestration project that will require the use of pore space located in more than one State.

(c) SDWA.—The term "SDWA" means the Safe Drinking Water Act.

(d) SUBSURFACE PROJECT BOUNDARY.—The term "subsurface project boundary" refers to the ex ante estimated spatial extent of free-phase injected CO_2, delineated by both the lateral and vertical boundaries from the time injection commences to the time that free-phase CO_2 ceases flowing, and taking into account a margin of error in predictions.

(e) UIC.—The term "UIC" means underground injection control.

(f) UIC CO_2 REGULATOR.—Except as provided in section 312(f) of this title, the term "UIC CO_2 Regulator" means a State that that has primary enforcement authority under section 1422 of the SDWA for the underground injection of carbon dioxide, and the Administrator of the Environmental Protection Agency for any other State.

(g) UNDERGROUND SOURCE OF DRINKING WATER.—The term "underground source of drinking water" means underground water with less than 10,000 mg/l total dissolved solids.

SUBTITLE B—PERMITS TO CONSTRUCT AND OPERATE GEOLOGIC SEQUESTRATION PROJECTS

SEC. 311. PURPOSE.

The purpose of this title is to expand and make more comprehensive the EPA's Class VI UIC permitting rules for the permanent geologic sequestration of carbon dioxide so that they balance the need for permanent geologic sequestration of CO_2 with the need to protect underground drinking water supplies so as to ensure the overall safety and protection of human health and the environment.

SEC. 312. PERMITTING UNDERGROUND INJECTION OF CARBON DIOXIDE FOR GEOLOGIC SEQUESTRATION.

(a) CO_2 UIC PROGRAM.—Not later than two years after the date of enactment of this Act, the Administrator shall revise and expand the Class VI UIC permitting rules for the injection of carbon dioxide for the purpose of geologic sequestration to include all requirements in this title, and take into consideration any recommendations made by the NRC CCS Technical Advisory Committee's first assessment carried out under section 202(b) of title II of this Act.

(b) PERMIT REQUIRED.—Except in accordance with a CO_2 injection permit issued by the UIC CO_2 Regulator under subsection (e), the following shall be unlawful:

(1) The construction of any injection well or underground facility for geologic sequestration, and

(2) The underground injection of carbon dioxide for geologic sequestration.

(c) NEW PROGRAM REQUIREMENTS.—The regulations under subsection (a) shall:

(1) provide that the UIC CO_2 Regulator shall issue a CO_2 injection permit in accordance with subsection (e) to any applicant that meets the requirements of this section;

(2) require, to the extent practicable, that the UIC CO_2 Regulator take an adaptive and performance-based approach to permitting the underground injection of carbon dioxide for permanent geologic sequestration as provided in title II of this Act;

(3) require that the applicant for the CO_2 injection permit satisfy the UIC CO_2 Regulator that the underground injection of carbon dioxide for permanent geologic sequestration will not endanger drinking water sources, unless the UIC CO_2 Regulator makes a SDWA applicability determination under subsection (d);

(4) require States seeking primary enforcement authority under the UIC program for permanent geologic sequestration of CO_2 to develop a framework and codify statutory authority for acquiring the necessary ownership interests, easements, or licenses necessary to occupy pore space;

(5) require State UIC programs to take into account the effects that permanent geologic sequestration of CO_2 in the permitting State will have in any other State; and,

(6) ensure that any State with a reasonable prospect of being affected by the grant of a CO_2 injection permit by another State shall have the right to intervene and participate in proceedings conducted by the permitting State for consideration of a petition of a permit for underground injection of CO_2 for permanent geologic sequestration.

(7) require the UIC CO_2 Regulator to develop and implement a communication strategy that maximizes transparency and educates the public about the permitting process and its enforcement that includes, but is not limited to:

(A) basic information about geologic sequestration, existing geologic sequestration projects, and its role in climate change mitigation, including a comparison to other low-carbon electric generation technologies;

(B) information about the permitting and risk mitigation processes, as well as about the enforcement of regulations over geologic sequestration project developers; and,

(C) information describing how members of the general public or the local host communities can voice concerns.

(8) require the geologic sequestration project developer to prepare a public engagement plan that includes but is not limited to:

(A) an assessment of the local community's needs, concerns, and any special social consideration that should be included in the engagement process;

(B) a plan to ensure that the community has easy access to information, including a strategy for the distribution of information about the project to the local community and stakeholders, including the media to be used and the timing of information releases during not only the permitting process but the active CO_2 injection and post-closure phases at the geologic sequestration site as well;

(C) a procedure for addressing community concerns;

(D) an overview of the information content that will be provided to the local community, addressing the costs, benefits and risks of the project, as well as mitigation, emergency management and contingency plans; and,

(E) a plan to reevaluate and revise the community engagement plan as necessary.

(9) require enhanced hydrocarbon recovery projects to hold a valid CO_2 injection permit in order to be considered a geologic sequestration project for the purposes of any Federal greenhouse gas emissions reduction program; and,

(10) establish a coordinated approach to certifying and permitting underground injection of CO_2 for permanent geologic sequestration, taking into account, and reducing redundancy with, all relevant statutory authorities.

(11) Regulations shall be subject to periodic review and revision as per title II, section 203(a), coordinating with, and taking into consideration the recommendations made by, the NRC CCS Technical Advisory Committee (title II, section 202(b)) and the Federal Geologic Sequestration Board (title II, section 203(b)).

(d) SDWA DETERMINATION.—

(1) DEFINITION—For purposes of this section, an SDWA determination is a determination that one or more provisions of Part C of the Safe Drinking Water Act are inapplicable in whole or in part to a specific source of underground drinking water affected by the CO_2 injection permit.

(2) DETERMINATION.—The UIC CO_2 Regulator may make an SDWA determination only if the regulator finds that the public benefit of geologic sequestration of carbon dioxide outweighs the protection of the underground drinking water source at issue after carefully balancing the goals of:

(A) minimizing the present and future threats to human health and the environment imposed by global climate change

With

(B) the protection and safety of underground drinking water sources.

(3) FACTORS.—In making an SDWA determination, the UIC CO_2 Regulator shall consider:

(A) direct and indirect impacts to underground sources of drinking water and human health and the environment resulting from geological sequestration of carbon dioxide.

(B) local impacts of potential surface leakage of sequestered carbon dioxide, assessing both probability and magnitude of potential harm,

(C) the nation's need to deploy and use CCS technology to control GHG emissions, and such other factors as the UIC CO_2 Regulator determines to be relevant.

(e) OPEN APPLICATION FOR AND ISSUANCE OF CO_2 INJECTION PERMITS.—

(1) If a geologic sequestration project developer applies for CO_2 injection permit under this subtitle, the UIC CO_2 Regulator shall:

(A) provide public notice in conformance with the requirements of 40 C.F.R. 25;

(B) conform to the procedures for decision-making under 40 C.F.R. 25; and,

(C) afford a period of 90 days for geologic sequestration project developers and operators who wish to contend that the grant of a CO_2 injection permit under this subtitle and, if applicable, a pore space permit under subtitle C may impair its own ability to develop and operate an alternative and potentially competing geologic sequestration project, or impair the operation of a currently operating geologic sequestration project, to intervene and file a competing application, such that:

(i) Interveners shall be entitled to equal consideration with the original applicant if it files a competing application within 90 days of public notice for the original application.

(ii) If a competing application is filed more than 90 days after the original application, the original application shall be considered and resolved upon its own merits without the necessity of consideration of the secondary competing application.

(2) The UIC CO_2 Regulator shall have the authority to attach to the issuance of the CO_2 injection permit, and to exercise the rights and privileges granted thereunder, such terms and conditions as are reasonable and necessary to effectuate the purposes of this title.

(3) The UIC CO_2 Regulator shall grant applications under this subsection upon a finding that the applicant is able and willing to properly do the acts and perform the service proposed in good faith, meet all relevant statutory and regulatory-imposed financial responsibilities and requirements, and conform to the requirements of this title, the rules thereunder, and the conditions of the CO_2 injection permit; otherwise, such application shall be denied.

(4) The UIC CO_2 Regulator may, after notice and opportunity for comment, revoke in whole or in part a UIC CO_2 injection permit issued under this section if the Regulator determines that the permit-holder has failed to comply with the re-

quirements of this title, the rules thereunder, or conditions of the permit.

(5) The EPA shall have authority to revoke or override any CO_2 injection permit issued by a State UIC CO_2 Regulator if:

(A) the State UIC CO_2 Regulator fails to take into account the effects that permanent geologic sequestration of CO_2 in the permitting State will have in any other State; and,

(B) the permanent geologic sequestration of CO_2 as authorized by the permit is determined to substantially endanger underground sources of drinking water (unless a SDWA determination is made under subsection (d)) and/or pose a threat to human health and the environment in neighboring States.

(f) INTERSTATE GEOLOGIC SEQUESTRATION PROJECTS.—

(1) States may enter into agreements with respect to permitting and regulating a geologic sequestration project that will require the use of geologic formations and pore space located in more than one State.

(2) The EPA is the UIC CO_2 Regulator for any interstate geologic sequestration project if the States where the project is located fail to enter into an agreement with respect to permitting and regulating the interstate project.

SEC. 313. CONFORMING AMENDMENTS.

(a) ENERGY INDEPENDENCE AND SECURITY ACT.—The first sentence of section 706 of title VII of the Energy Independence and Security Act (42 U.S.C. § 17254) is amended—

(1) by striking out "The" at the beginning of the first sentence and inserting "Subject to title III of the Carbon Capture and Sequestration Regulatory Act of 2012, the"; and,

(2) by striking out "Nothing" at the beginning of the second sentence and inserting "Subject to such title, nothing".

(b) SAFE DRINKING WATER ACT.—

(1) Section 1421(b)(1) of the Safe Drinking Water Act of 1974 as amended (42 U.S.C. § 300h) is amended by:

(A) inserting at the beginning of subparagraph (B) "except as provided in section 312(d) of the Carbon Capture and Sequestration Regulatory Act of 2012," and,

(B) inserting a new subparagraph "(E)", which shall read "Shall meet all requirements of the regulations in effect under title III of the Carbon Capture and Sequestration Regulatory Act of 2012."

(2) Section 1422(b)(1)(A)(i) of the Safe Drinking Water Act (42 U.S.C. § 300h-1) is amended by inserting at the end thereof "and title III of the Carbon Capture and Sequestration Regulatory Act of 2012."

SUBTITLE C—ACCESS TO PORE SPACE FOR GEOLOGIC SEQUESTRATION

SEC. 321. PURPOSE.

The purpose of this subtitle is to establish a fair, equitable, and elective permitting procedure for the allocation, management, and use of subsurface pore space for permanent geologic sequestration of carbon dioxide, thereby reducing carbon dioxide emissions to the atmosphere; and, to the maximum extent practicable, protecting private property interests and preventing subsurface property disputes from arising.

SEC. 322. DEFINITIONS.

For the purpose of this subtitle:

(a) MATERIAL IMPAIRMENT.—The term "material impairment" means the subsurface interest-holder has suffered actual and substantial damages resulting from the injection or migration of carbon dioxide.

(b) NON-SPECULATIVE ECONOMIC INTEREST.—The term "non-speculative economic interest" means the ability to recover actual mineral resources or engage in other current or imminent subsurface activities that have substantial economic value. It shall be presumed, subject to rebuttal, that use of pore space for which a CO_2 injection permit is required under subtitle B is a speculative interest until such a permit is issued. This presumption shall be overcome if the ownership interests, easements, or licenses necessary to use the pore space for permanent geologic sequestration of CO_2 have been acquired by the project developer through either State statutory authority or voluntary contract *and* the geologic strata containing the pore space is actively being characterized and tested for the purposes of developing a geologic sequestration project.

(c) PORE SPACE PERMIT.—The term "pore space permit" refers to a permit issued by the UIC CO_2 Regulator under subtitle C authorizing the exclusive right to access and use pore space for the permanent geologic sequestration of CO_2 within the subsurface project boundary.

(d) PREEXISTING INTEREST.—The term "preexisting interest" means an interest in demonstrated economically recoverable mineral resources or in other subsurface activities that are non-speculative economic interests.

(e) SUBSURFACE TRESPASS.—The term "subsurface trespass" means geologic sequestration by a site operator that results in a physical invasion of pore space in which the site operator does not have the requisite ownership interest, easement, or license to occupy.

SEC. 323. MANAGING ACCESS AND USE OF PORE SPACE FOR PERMANENT GEOLOGIC SEQUESTRATION OF CARBON DIOXIDE.

(a) EFFECT OF CO_2 INJECTION PERMIT.—

(1) If a CO_2 injection permit issued to a site operator under subtitle B is in effect, no permit may be issued to any other person to inject CO_2 for the purposes of geologic sequestration within the subsurface project boundary defined in the existing permit, unless the existing permit provides otherwise.

(2) A CO_2 injection permit under subtitle B does not relieve the site operator of any liability for failure to obtain ownership interests, easements, or licenses necessary to occupy pore space unless the CO_2 injection permit incorporates a pore space permit issued under this section.

(b) REGULATIONS.—Not later than one year after the date of enactment of this Act, the Administrator shall promulgate rules and procedures for allocating and managing the use of subsurface pore space for the purpose of permanent geologic sequestration of CO_2, and shall integrate the administration of these rules and procedures with existing State and Federal UIC programs such that if a pore space permit is issued under this section in connection with a CO_2 injection permit, the pore space permit conveys the exclusive Federal privilege to access and use of pore space for geologic sequestration of carbon dioxide within the subsurface project boundary defined by the permit.

(c) APPLICATION FOR PORE SPACE PERMIT.—If a CO_2 injection permit applicant seeks to develop a geologic sequestration project in a State (or States) that has (or have) not codified statutory authority for acquiring and using pore space for permanent geologic sequestration of CO_2 *and* the applicant has not acquired through voluntary contract the ownership interests, easements, or licenses necessary to occupy pore space, the project developer, in connection with an application for a CO_2 injection permit under subtitle B, may include a request for a pore space permit.

(1) If a request is made for a pore space permit, the UIC CO_2 Regulator shall:

(A) provide public notice in conformance with the requirements of 40 C.F.R. 124.10, and

(B) afford a period of 60 days for participation in the permit application proceeding to all interested parties and holders of preexisting interests that would be materially impaired by the granting of the pore space permit.

(2) If an interested party or the holder of a preexisting interest fails to intervene in the pore space permit application not later than the date of filing of notice paragraph (1)(A):

(A) Except as provided in subparagraph (B), such party shall be deemed to have waived any and all rights and property interests that become impaired by the project that is the subject of the proceeding should a pore space permit be issued by the UIC CO_2 Regulator.

(B) An interested party or the holder of a preexisting interest may be permitted late intervention in a proceeding under this section upon a showing of good cause.

(3) A competing geologic sequestration project applicant who intervenes under paragraph (1)(B) must indicate their intention to file a competing CO_2 injection permit application within 90 days of the original project application in accordance with section 312(e)(1)(C) of subtitle B.

(4) A pore space permit shall not be issued if the applicant has not already been issued a CO_2 injection permit under section 312(e) of this title, or is not concurrently applying for both a CO_2 injection permit and a pore space permit.

(d) Consideration Of Preexisting Interests In Scope Of Project.—

(1) If it is demonstrated that a preexisting interest would be materially impaired by the granting of a CO_2 injection permit for permanent geologic sequestration of carbon dioxide, the

geologic sequestration project should be permitted only in accordance with:

(A) a modification of the project that avoids the impairment; or

(B) a contractual agreement between the owner of the preexisting interest and the project applicant; or

(C) a finding by the UIC CO_2 Regulator that condemnation of the preexisting interest through the exercise of eminent domain pursuant to subsection (e) of this subtitle, with appropriate compensation, is necessary to the proper operation the geologic sequestration project under application.

(2) In connection with a pending application proceeding for geologic sequestration of carbon dioxide under this subsection, the UIC CO_2 Regulator may exclude from the subsurface project boundary, or authorize the exercise of eminent domain under subsection (e) for, any portion of a geologic formation where the formation is:

(A) subject to active and properly licensed exploration or production of hydrocarbon resources or hard minerals;

(B) actively used for the properly licensed injection of brines or other fluids for the purpose of enhanced recovery of hydrocarbon resources;

(C) actively used for storage of crude oil;

(D) actively used for injection of fluid wastes for disposal pursuant to a valid UIC permit;

(E) actively used for certificated natural gas storage;

(F) actively used for properly licensed groundwater recharge and recovery;

(G) actively used for properly licensed compressed air storage;

(H) actively used for geothermal electric power generation;

(I) actively being subjected to geophysical and environmental testing for the purpose of developing a geologic sequestration project, provided that the ownership interests, easements, or licenses necessary to occupy the pore space for permanent geologic sequestration of CO_2 have been acquired by the project developer either through State statutory authority or voluntary contract; or

(J) actively used for other purposes the UIC CO_2 Regulator deems relevant.

(e) Right Of Eminent Domain.—When the UIC CO_2 Regulator finds that condemnation of a preexisting interest is necessary to ensure the proper operation and protect integrity of the geologic sequestration project under application:

(1) the holder of a CO_2 injection permit to whom a pore space permit was issued under subsection (c) of this subtitle, when it cannot acquire by contract, or is unable to agree with the owner of the preexisting interest to the compensation to be paid for the necessary surface and subsurface property rights to construct, operate, and maintain underground wells and facilities for the permanent sequestration of carbon dioxide, may acquire the same by the exercise of the right of eminent domain in the district court of the United States for the district in which such property interest is located; and,

(2) the practices and procedures in any action or proceeding for that purpose in the United States shall conform as nearly as possible with the practices and procedures in a similar action or

proceeding in the courts of the State where the property is situated.

(f) DOMINANCE OF MINERAL ESTATE.—

(1) The provisions of this subtitle shall not be deemed to preempt the mineral rights laws of any State, except to the extent necessary to ensure that mineral exploration and production activities will not cause leakage of permanently sequestered carbon dioxide, or compromise the integrity of the geologic sequestration site.

(2) The holder of a State-law right to conduct mineral exploration or production activities shall not be entitled to compensation as a result of any such activities being precluded or restricted, to the extent necessary, to protect the integrity of the geologic sequestration site.

(3) As with all other property interests, mineral rights are subject to condemnation through the exercise of eminent domain under this subtitle.

(g) FEDERAL REMEDY FOR CLAIMS OF SUBSURFACE TRESPASS.—

(1) A claim of subsurface trespass shall not be actionable against a site operator conducting geologic sequestration in accordance with a valid CO_2 injection permit issued by the UIC CO_2 Regulator and to whom a pore space permit has been issued under subsection (c) of this subtitle unless the injection or migration of carbon dioxide materially impairs:

(A) preexisting interests that were identified to the UIC CO_2 Regulator during permit proceeding pursuant to subsections (c) and (d) of this subtitle; or

(B) interests outside the subsurface project boundary.

(2) The issuance of a CO_2 injection permit shall not protect a site operator from claims of subsurface trespass if the injection or migration of carbon dioxide materially impairs preexisting interests established during the permit proceeding that have not been compensated via a contractual agreement between the owners of the preexisting interests and the project applicant, or condemned through the valid exercise of eminent domain pursuant to subsection (e) of this subtitle.

(3) A surface or subsurface property interest-holder shall be permitted to recover money damages only for loss of a non-speculative value resulting from the injection and migration of carbon dioxide.

(A) The standard for calculating money damages shall be the present value of the demonstrated impairment, or the otherwise expected value of the future income stream that would have accrued had the interest not been impaired.

(B) Punitive damages shall be barred if the site operator who causes the material impairment acts in compliance with the terms of the CO_2 injection permit.

(C) Any damage award shall be discounted by the cost of the mineral extraction or current and actual subsurface activity that is not a result of impairments caused by the injection and migration of carbon dioxide.

(4) Injunctive relief for subsurface trespass shall not be allowed unless the holder of the property interest shows that the harm to the property interest clearly outweighs the utility of the sequestration of carbon dioxide.

(5) The United States district court for the district in which a trespass claim arises shall have exclusive jurisdiction over such a claim, and the United States Court of Appeals for the

District of Columbia shall hear any appeal of a district court ruling under this subsection.

(h) REGULATORY TAKINGS.—

(1) Any claim for a regulatory taking without just compensation that arises out of an action by the United States or a State under this title shall be filed against the United States in the United States Court of Federal Claims pursuant to chapter 91 of title 28 of the United States Code.

(2) No such claim may be filed against a State by reason of its action as a UIC CO_2 Regulator in accordance with this title.

SUBTITLE D—GEOLOGIC SEQUESTRATION OF CARBON DIOXIDE ON FEDERAL LANDS

SEC. 331. PURPOSE.—

The purpose of this section is to expressly authorize the Secretary of the Interior, through the Bureau of Land Management, to license the use of Federal lands for the permanent geologic sequestration of carbon dioxide.

SEC. 332. DEFINITIONS.

For the purposes of this subtitle:

(a) BLM.—The term "BLM" means the Bureau of Land Management.

(b) FEDERAL LANDS.—The term "Federal lands" refers to lands managed by the BLM and FS that have been determined by each agency to be available for use for permanent geologic sequestration of carbon dioxide.

(c) FS.—The term "FS" means the Forestry Service.

(d) PLAN OF DEVELOPMENT.—The term "Plan of Development" means a detailed description of the design of the geologic sequestration project and its facilities submitted by a project development license applicant to the BLM. The Plan of Development helps the BLM assess the public safety and environmental effects of the proposed geologic sequestration project and its facilities.

(e) SECRETARY.—Unless otherwise specified, the term "Secretary" means the Secretary of the United States Department of the Interior.

(f) SPLIT ESTATE.—The term "split estate" means lands in which the Federal government owns the surface rights, but the mineral rights are privately owned.

SEC. 333. PERMITTING AND LICENSING GEOLOGIC SEQUESTRATION ON FEDERAL LANDS.

(a) REGULATIONS.—Not later than one year after the date of enactment of this title, the Secretary shall promulgate rules and procedures for allocating and managing the use of deep subsurface pore space for the purpose of geologic sequestration of carbon dioxide on Federal lands and split estates.

(b) REQUIREMENTS.—

(1) Through an integrated permitting system, the BLM and the EPA should jointly license sequestration projects on Federal lands and split pursuant to the UIC permitting procedures and requirements under subtitle B of this title.

(2) TWO-PHASE APPLICATION PROCEDURE.—The BLM shall issue perpetual easements under title V of the Federal Land Management and Policy Act for geologic sequestration exploration and development projects on Federal lands through a two-phase license procedure.

(A) PHASE I: PROJECT AREA LICENSE.—A project area license grants the geologic sequestration project applicant the exclusive right to conduct environmental and geological testing and monitoring within the proposed subsurface project boundary.

(i) Priority of application will be given to the first complete application the BLM receives.

(ii) The project area license term shall be six years.

(iii) During the six-year project area license term, the pore space within the proposed subsurface project boundary shall not be available for other geologic sequestration project easements.

(iv) The holder of a project area license shall formulate a Plan of Development during the six-year license term.

(B) PHASE II: PROJECT DEVELOPMENT LICENSE.—A project development license, in combination with a CO_2 injection permit issued pursuant to subtitle B, grants the project applicant the right to construct, operate, and maintain underground wells as well as surface and subsurface facilities for the permanent geologic sequestration of CO_2 within the subsurface project boundary defined in the CO_2 injection permit and project development license.

(i) A project development license shall be awarded based on the merits of the applicant's Plan of Development.

(ii) No CO_2 injection permit or project development license may be issued to any other person to inject CO_2 for the purpose of geologic sequestration of CO_2 within the subsurface project boundary defined in the existing CO_2 injection permit and project development license.

(3) LICENSING GEOLOGIC SEQUESTRATION PROJECTS ON SPLIT ESTATES.—The EPA CO_2 injection permit proceedings conducted under subtitle B for geologic sequestration projects

proposed on split estates shall be subject to the requirements under subtitle C of this title if the project applicant has not acquired the ownership interests, easements, or licenses necessary to occupy pore space in all applicable states through state statutory authority or voluntary contract, and instead elects to apply for a pore space permit pursuant to subtitle C.

(4) COLLECTION OF ANNUAL FEES AND RENT.—The BLM shall annually charge administrative fees and collect an annual nominal rent from project operators for the administration and management of geologic sequestration licenses.

(5) ENVIRONMENTAL REVIEW OF FEDERAL ACTIONS AUTHORIZING PRIVATE GEOLOGIC SEQUESTRATION PROJECTS ON FEDERAL LANDS AND SPLIT ESTATES.—The BLM shall be the lead agency for the purpose of Federal environmental review pursuant to the National Environmental Policy Act of 1969 (42 U.S.C. § 4321 et seq.) and, in consultation with all affected agencies, shall prepare a single environmental review document for each license application phase to be used as the basis for decisions under Federal law related to each project phase.

(A) NEPA REVIEW OF THE PROJECT AREA LICENSE.—The environmental review for the project area license shall only assess likely environmental effects of the testing facilities to be constructed and operated under the terms of that license. The environmental effects of permanent geologic sequestration shall not be considered for environmental review of the project area license.

(B) NEPA REVIEW OF THE PROJECT DEVELOPMENT LICENSE.—The environmental review for the project development license shall assess the likely environmental effects of permanent geologic sequestration of CO_2 and the facilities and wells constructed for such purpose.

SEC. 334. CONFORMING AMENDMENTS.

(a) Section 302 (b) of the Federal Land Policy and Management Act of 1976 (43 U.S.C. § 1732) is amended by inserting after "utilize public lands for habitation, cultivation," the following:

(1) "permanent geologic sequestration of carbon dioxide,".

(b) Section 501 (a) of the Federal Land Policy and Management Act of 1976 as amended (43 U.S.C. § 1761) is amended by adding a new paragraph "(a)(8)" and inserting "pipelines and other systems for the transportation or distribution of carbon dioxide and for the permanent geologic sequestration of carbon dioxide and terminal facilities in connection therewith."

TITLE IV—LONG-TERM STEWARDSHIP OF CLOSED INJECTION SITES

SEC. 401. DEFINITIONS.

For purposes of this title:

(a) APPLICABLE REGULATORY AUTHORITY.—The term "applicable regulatory authority" means the Federal or State agency that has principal responsibility for issuing injection permits for, and regulating underground injection by, a GS project.

(b) CERTIFICATE OF CLOSURE.—The term "certificate of closure" means a determination issued by the applicable regulatory authority with respect to a GS project that certifies that the operator of the project has completed injection operations, well closure, and any required monitoring and remediation to ensure that any carbon dioxide injected into a geologic formation would not harm or present a risk to human health, safety, and the environment, including drinking water supplies.

(c) CIVIL CLAIM.—The term "civil claim" means a claim before any court or administrative agency for civil relief with respect to damage from the injection of carbon dioxide by a GS project.

(d) DAMAGE.—

(1) IN GENERAL.—The term "damage" means any direct or indirect damage or harm to persons, property, or natural resources from the injection of carbon dioxide into a geologic formation.

(2) INCLUSIONS.—The term "damage" includes personal injury, sickness, real or personal property damage, natural resource damage, trespass, subsidence losses, revenue losses, and loss of profits.

(e) ENHANCED HYDROCARBON RECOVERY.—The term "enhanced hydrocarbon recovery" means the use of carbon dioxide to improve or enhance the recovery of oil or natural gas from oil or natural gas fields.

(f) FUND.—The term "Fund" means the Carbon Sequestration Trust Fund established by section 403(d)(1).

(g) FIRST MOVER PROJECT.—The term "first mover project" refers to a sequestration project determined by the Secretary to be one of the first of its kind to be operated and for this reason warranting special attention if deemed to be in the national interest.

SEC. 402. LONG-TERM STEWARDSHIP RESPONSIBILITY.

(a) IN GENERAL.—Except as otherwise provided in subsection (b), the Federal Geologic Sequestration Board shall be responsible for the long-term stewardship of a GS project at such time as it issues a certificate of closure for the project.

(b) TRANSFER TO STATE.—

(1) IN GENERAL.—A State may request that the management responsibilities associated with long-term stewardship of a GS project located in the State be transferred to the State in accordance with regulations prescribed by the Board.

(2) APPROVAL OF REQUEST.—If the Board approves a request under paragraph (1), the State shall be responsible for the long-term stewardship of the applicable GS project beginning on the date of the approval. The Secretary is authorized to enter into contracts to reimburse States for their expenses of stewardship.

(3) FAILURE TO ACT BY STATE.—In accordance with any regulations under paragraph (1), if the Board determines that a State that has accepted management responsibilities under paragraph (1) has failed to carry out the responsibilities of the State with respect to the GS project, the Board shall assume long-term stewardship of the project as soon as practicable after the date of the determination.

SEC. 403. LONG-TERM STEWARDSHIP PROGRAM.

(a) IN GENERAL.—The Board shall administer the long-term stewardship program to be established by this title.

(b) STEWARDSHIP FEES.—In carrying out the stewardship program under this title, the Board shall require operators of GS projects to pay a risk-based fee at the time of injection, in an amount to be established in accordance with subsection (c), for each ton of carbon dioxide injected by the GS project into geologic formations during the operational phase of the facility.

(c) AMOUNT OF FEES.—

(1) SETTING FEES.—

(A) IN GENERAL.—As soon as practicable after the date of enactment of this Act, the Board shall establish, and may from time-to-time revise—

(i) the minimum and maximum balance for the Fund; and,

(ii) the amount of the fee required under subsection (b).

(B) CRITERIA.—In establishing fees under subparagraph (A)(ii), the Board shall take into account—

(i) the estimated quantity of carbon dioxide to be injected annually into geologic formations by all operating commercial GS projects;

(ii) the risk of an incident resulting in liability;

(iii) the likely dollar value of any damages relating to an incident;

(iv) other factors relating to the risk of the GS project and associated geologic formation; and,

(v) impact on commercial and economic viability of GS projects.

(C) CONSIDERATION OF PROJECT RISKS.—In establishing the amount of the fees under subparagraph (A)(ii), the Board shall, to the extent practicable, use a fee system that is based on the level of risk associated with a specific geologic formation and project development plan and operator history, in order to provide an incentive for the selection and operation of the well-performing GS projects.

(D) REVIEW AND ADJUSTMENT.—The Board shall, on at least an annual basis, review the Fund balance—

(i) to ensure that there are sufficient amounts in the Fund to make the payments required under subsection (d)(2)(A); and,

(ii) to determine whether or not to increase or decrease the amount, or discontinue collection, of the fee, after taking into consideration—

(I) the annual quantity of carbon dioxide injected by carbon dioxide storage facilities;

(II) the number and estimated value of current and projected claims against the Fund; and,

(III) any other relevant factors, as determined by the Board.

(2) DEPOSIT.—Notwithstanding section 3302 of section 31, United States Code, the fees collected under paragraph (1) shall be deposited in the Fund.

(d) CARBON SEQUESTRATION TRUST FUND.—

(1) ESTABLISHMENT.—There is established in the Treasury of the United States a revolving fund, to be known as the "Carbon Sequestration Trust Fund", consisting of such amounts as are deposited under this section.

(2) USE OF FUND.—

(A) IN GENERAL.—Amounts in the Fund shall be made available, without further appropriation or fiscal year limitation—

(i) to the Board for the payment, in accordance with section 404, of civil claims respecting a GS project that are brought after a certificate of closure for the project has been issued;

(ii) to the Board, or State agency with Board approval, for long-term stewardship after the date of issuance of a certificate for closure;

(iii) to the Board, or State agency with Board approval, to pay any reasonable and verified administrative cost incurred by the Board or State agency in carrying out the stewardship program;

(iv) to the Board for emergency remediation actions under section 407; and,

(v) to the Board, for the cost of any compensatory action that may be required under title V.

(B) LIMITATION.—Amounts in the Fund shall only be used for the purposes described in clause (i), (ii), (iii), or (iv) or (v) of subparagraph (A).

(3) REPAYABLE ADVANCES.—

(A) IN GENERAL.—If sufficient amounts are not available in the Fund to cover potential claims, the Board may request from the Secretary of the Treasury an interest-bearing advance in funding from the Treasury to carry out the Program.

(B) TERMS AND CONDITIONS.—The terms and conditions for the repayment of an advance under subparagraph (A) shall be specified by the Secretary of the Treasury.

SEC. 404. CIVIL CLAIMS.

(a) BOARD ADJUDICATION OF CLAIMS.—The Board shall pay any civil claim for damage respecting a GS project for which a certificate of closure has been issued. Such claim shall be—

(1) filed with the Board;

(2) adjudicated in accordance with sections 554, 557, and 558 of title 5, United States Code; and,

(3) paid in accordance with compensation rules under 405(b)(3) of this title.

(b) LIABILITY OF OTHER ENTITIES.—After issuance of a certificate of closure to a GS project, a civil claim respecting such project may not be brought against—

(1) the owner or operator of the GS project, except as provided in subsection (c);

(2) the generator of the carbon dioxide injected into the applicable geologic formation in connection with the GS project;

(3) the owner or operator of the pipeline used to transport the carbon dioxide to the GS project; or

(4) any State agency that has long-term stewardship responsibility under section 402(b).

(c) BOARD RECOVERY FROM OPERATOR.—Notwithstanding subsection (b)(1), the operator of a GS project shall be liable to the Board for any compensation paid or remediation cost incurred as a result of—

(1) the project operator's willful failure to comply with any regulatory requirement;

(2) the reliance by the Federal or State agency issuing a certificate of closure on any untrue statement of a material fact by the project operator; or

(3) the omission by the project operator of a statement of material fact necessary in order to make the statements made

not misleading (in light of the circumstances under which they were made) to the Federal or State agency.

(d) CLAIMS PRIOR TO CERTIFICATE OF CLOSURE.—Nothing in this title affects civil liability for damage claims asserted prior to issuance of a certificate of closure.

SEC. 405. FEDERAL GEOLOGIC SEQUESTRATION BOARD.

(a) ESTABLISHMENT.—There is hereby established, within the Department of Energy, an independent agency to be known as the Federal Geologic Sequestration Board. The Board shall consist of three members appointed by the President and with the advice and consent of the Senate for terms of six years, except that—

(1) the members first appointed shall serve for terms of two, four, and six years as designated by the President at the time of appointment, and

(2) a member appointed to fill an unexpired term shall serve only for the remainder of that term.

Members of the Board shall be compensated at the rate prescribed for Level IV of the Executive Schedule.

(b) FUNCTIONS.—

(1) IN GENERAL.—The Board shall carry out the duties prescribed for it under this title; and exercise such other authorities as may be necessary or appropriate to carry out its functions under this title, including employment of personnel and entering into contracts.

(2) STANDARDS.—The Board, after consultation with the Administrator, shall establish standards for any monitoring, measurement, verification, and site remediation activities necessary to protect health, safety, and the environment during long-term stewardship.

(3) COMPENSATION STANDARDS, PROCEDURES, AND SCHEDULES.—The Board shall by rule prescribe compensation

standards, procedures, and schedules for determining the nature and amount of compensation that will be paid from the Trust Fund for civil claims for damage by the Board under this title. Rules written under this paragraph shall include procedures under which the Board will take any compensatory action required under section 503.

(4) PRIVATE INSURERS.—The Board may contract with private insurers to provide claim adjustment services for public liability claims.

(c) POWERS.—The Board has the authority to—

(1) prescribe, by rule or order, such requirements for monitoring GS projects and for making such inspections and reports as may be necessary or appropriate to carry out this title;

(2) enter onto the premises or property of any GS project to carry out this title;

(3) commence a civil action in the United States District Court to recover from any project operator any fees not paid when due, after notice and an opportunity to cure any deficiency within thirty days of such notice;

(4) bring an action against any person in the United States District Court to enforce the provisions of this title or rules or orders thereunder, and to obtain appropriate injunctive or other relief; and,

(5) seek civil penalties for violations of provisions of this Act, as provided under subsection (d).

(d) PENALTIES.—

(1) CIVIL PENALTIES.—Any person who knowingly violates any provision of this title or any rule or order thereunder shall be subject to a civil penalty of $50,000 per violation.

(2) CRIMINAL PENALTIES.—Any person who knowingly and willfully violates any provision of this title or any rule or

order thereunder shall be subject to a fine of $500,000 or imprisonment for a term of two years, or both.

(e) JUDICIAL REVIEW.—Rules of general applicability prescribed under this title by the Board shall be reviewed by the United States Courts of Appeal in accordance with chapter 158 of title 28, United States Code. All other agency actions under this title shall be reviewed in accordance with chapter 7 of title 5, United States Code.

SEC. 406. FIRST MOVER PROJECTS.

(a) INDEMNIFICATION.—Notwithstanding section 1341 of title 31, United States Code, but subject to limitations in Appropriations Acts, the Secretary is authorized to enter into agreements to indemnify owners and operators of qualifying geologic sequestration projects for all or part of the costs incurred to satisfy civil claims for damage (whenever made) that arise from carbon dioxide injection into a geologic formation during the demonstration period of a first mover project, as determined by the Secretary. The Secretary may impose such conditions on such indemnification agreements as may be necessary or appropriate to protect the financial interest of the United States, including a requirement that indemnification provided to qualifying projects under this section be limited to the extent that the Secretary determines that potential long-term liabilities can be adequately addressed through the implementation of preceding sections of this title.

(b) QUALIFYING PROJECTS.—A project shall qualify for indemnification under this section only if the project—

(1) receives Federal funds to demonstrate the geologic sequestration of carbon dioxide at a GS project;

(2) obtains all necessary permits for the injection of carbon dioxide into a suitable geologic formation;

(3) injects the carbon dioxide into geologic formations that do not involve enhanced oil or gas recovery; and,

(4) agrees to comply with conditions that the Secretary may establish to protect health, environment and safety, including requirements on volume of injection, depth of injection, purity of injectate, and proximity of underground drinking water sources, human settlements, or ecologically sensitive areas.

SEC. 407. EMERGENCY SITE REMEDIATION.

The Administrator may undertake emergency remediation action at any GS project for which a certificate of closure has not been issued if in the judgment of the Administrator an imminent danger to public health or safety so requires. Amounts held in the Fund shall be available for such purpose. The owner and the operator of the GS project shall be liable to the Administrator for any costs incurred for such remediation action. In addition the Board shall consider the cost of such expected or actual remediation when setting fees under section 403(c)(1)(C). The Administrator shall maintain, directly or by prior contract, a stand-by emergency response capability for purposes of this section, and may enter into both advance and responsive contracts to carry out his responsibilities under this section.

TITLE V—ACCOUNTING FOR SEQUESTERED CARBON DIOXIDE

SEC. 501. PURPOSE; DEFINITIONS.

(a) PURPOSE.—The purpose of this title is to:

(1) establish inventory accounting and reporting procedures for CO_2 capture facilities, CO_2 pipelines, and geologic sequestration facilities, that will allow assessment of the efficacy of the overall CCS system;

(2) establish leakage monitoring and reporting procedures for geologic sequestration facilities to be used in the event that injected CO_2 leaks back to the atmosphere; and,

(3) specify the use of these measures in GHG emission reduction programs.

(b) DEFINITIONS.—For the purposes of this title:

(1) COMPENSATORY ACTION.—The term "compensatory action" refers to any requirement to submit allowances or to take other action to compensate for leakage in accordance with the provisions of a Federal greenhouse gas emission reduction program.

(2) LEAKAGE.—The term "leakage" refers to the release to the atmosphere of carbon dioxide that was injected for purposes of geologic sequestration.

SEC. 502. ACCOUNTING FOR CO_2 HANDLED IN CCS.

(a) DEVELOPMENT OF STANDARDS.—

(1) Not later than one year after the date of enactment of this Act, the Administrator shall:

(A) establish inventory accounting standards for the measurement of CO_2 handled by CO_2 capture facilities, CO_2 pipelines, and geologic sequestration facilities considering:

(i) current standards established by industry bodies;

(ii) best practices in the process industries;

(iii) advice of the CCS Technical Advisory Committee of the National Research Council; and,

(iv) compatibility with Federal greenhouse gas emissions reporting programs.

(B) review and revise leakage monitoring and reporting requirements for geologic sequestration facilities established in 40 C.F.R. Part 98 Subpart RR in order to:

(i) incorporate the lessons learned from demonstration geologic sequestration facilities;

(ii) identify the level of leakage a monitoring program must be designed to detect; and,

(iii) require focused surface measurements to locate and quantify leakage emissions only if routine subsurface monitoring, as required in title III and this title, indicates loss of containment, and near-surface monitoring indicates that leakage is occurring.

(C) establish a national reporting system that will allow annual estimates of the mass of CO_2 captured, transported, injected, vented and reemitted as leakage, as well as the mass of any CO_2 lost or unaccounted for in handling.

(b) MEASUREMENT AND REPORTING REQUIREMENTS.—According to the standards established in subsection (a), the following entities must measure and report the following information:

(1) CO_2 CAPTURE FACILITY.—A CO_2 capture facility must measure and annually report the mass of CO_2 captured and the amount exported for sequestration or any other purpose in accordance with standards established by the Administrator.

(2) CO_2 PIPELINE.—A project transporting CO_2 in whole or part for the purposes of geologic sequestration must measure and annually report the mass of CO_2 imported from capture facilities and the amount exported to geologic sequestration facilities, in accordance with standards established by the Administrator.

(3) GEOLOGIC SEQUESTRATION FACILITY.—A geologic sequestration facility must measure and annually report the following information:

(A) the amount of CO_2 imported, and the amounts vented and injected, in accordance with standards established by the Administrator;

(B) leakage quantity, as measured through monitoring in accordance the standards established in subsection (a); and,

(C) the net quantity of CO_2 sequestered, which shall be determined as the difference between the quantity injected, and the leakage quantity, reported pursuant to paragraphs (A) and (B) of this subsection.

(c) ANNUAL RECONCILIATION.—

(1) On an annual basis, the Administrator will:

(A) Reconcile the mass of CO_2 captured, transported, vented, injected, and emitted as leakage by CO_2 capture facilities, CO_2 pipelines, and geologic sequestration facilities in order to determine:

(i) the amount of CO_2 lost or unaccounted for in handling; and

(ii) the parties responsible for CO_2 lost or unaccounted above a threshold quantity to be determined by the Administrator.

(B) Publish in the Federal Register and on the Agency's website at a level of detail to be determined by the Administrator:

(i) the mass of CO_2 captured, transported, injected, and emitted to the atmosphere by CO_2 capture facilities, CO_2 pipelines, and geologic sequestration facilities; and,

(ii) the mass of CO_2 otherwise unaccounted for.

SEC. 503. INTEGRATION WITH GHG EMISSION REDUCTION PROGRAM

(a) APPLICABILITY.—This section applies only if a Federal greenhouse gas emission reduction program (1) is in effect, (2) is applicable to stationary sources, and (3) provides that carbon dioxide injected for geologic sequestration is eligible for or subject to an otherwise applicable compliance obligation under Federal law.

(b) ELIGIBILITY OF GEOLOGIC SEQUESTRATION FACILITIES.—Only carbon dioxide injected at a site holding a valid CO_2 injection permit will be considered geologically sequestered under any Federal greenhouse gas emission reduction program.

(c) DEVELOPMENT OF STANDARDS FOR BACKSTOP COMPENSATORY ACTION.—Not less than one year after the enactment of a Federal greenhouse gas emission reduction program, and after the report of the CCS Technical Advisory Committee to the National Research Council, the Administrator shall promulgate regulations that specify the requirements for backstop compensatory action for potential leakage from sites where subsurface monitoring indicates loss of containment, but a satisfactory surface monitoring program, as described in section 502(a)(1), cannot be implemented at the site due to technological or other factors. Backstop compensatory action shall be determined as a fraction of the total amount of CO_2 sequestered at the site.

(d) COMPENSATORY ACTION FOR CARBON DIOXIDE LOST OR UNACCOUNTED FOR BEFORE INJECTION.—

(1) Any CO_2 capture facility, CO_2 pipeline, or geologic sequestration facility (or combination thereof) that the Administrator determines is responsible for a material quantity of lost or unaccounted for CO_2, in accordance with the inventory reconciliation specified in section 502(c), must take appropriate compensatory action with respect to such lost or unaccounted for CO_2, in accordance with standards established by the Administrator, consistent with applicable provisions of a greenhouse gas emission reduction program.

(e) COMPENSATORY ACTION FOR LEAKAGE OF CARBON DIOXIDE FROM GEOLOGIC SEQUESTRATION SITES.—

(1) MEASURED LEAKAGE.—Operators of geologic sequestration facilities shall be required to take appropriate compensatory action with respect to any leakage identified and reported in accordance with regulations established in section 502 and the provisions of the greenhouse gas emission reduction program.

(2) POTENTIAL LEAKAGE.—If subsurface monitoring indicates loss of containment, but the Administrator determines that a satisfactory surface monitoring program, as described in section 502(a), cannot be implemented at the site due to technological or other factors, the operator must take backstop compensatory action, in accordance with regulations established in section 503(c). Backstop compensatory action must be taken annually until accurate surface monitoring is completed or subsurface monitoring indicates that injected CO_2 is successfully contained in a geologic formation.

NOTES

1 THE IMPORTANCE OF CARBON CAPTURE AND GEOLOGIC SEQUESTRATION IN A CARBON CONSTRAINED WORLD

1 Intergovernmental Panel on Climate Change (IPCC). 2008. *Climate Change 2007: Synthesis Report. Contribution of Working Groups I, II and III to the Fourth Assessment Report of the Intergovernmental Panel on Climate Change (Core Writing Team, Pachauri, R. K. and Reisinger, A.(eds.)).* http://www.ipcc.ch/publications_and_data/publications_ipcc_fourth_assessment_report_synthesis_report.htm (accessed December 26, 2010).

2 Northeast Climate Impacts Assessment (NECIA). 2006. *Climate Change in the U.S. Northeast, A Report of the Northeast Climate Impacts Assessment, Union of Concerned Scientists.* http://www.northeastclimateimpacts.org (accessed December 26, 2010).

3 Hoegh-Guldberg, O., P. J. Mumby, A. J. Hooten, R. S. Steneck, P. Greenfield, E. Gomez, C. D. Harvell, P. F. Sale, A. J. Edwards, K. Caldeira, N. Knowlton, C. M. Eakin, R. Iglesias-Prieto, N. Muthiga, R. H. Bradbury, A. Dubi, and M. E. Hatziolos. 2007. Coral reefs under rapid climate change and ocean acidification. *Science* 318: 1737–1742.

4 Orr, James C., Victoria J. Fabry, Olivier Aumont, Laurent Bopp, Scott C. Doney, Richard A. Feely, Anand Gnanadesikan, Nicolas Gruber, Akio Ishida, Fortunat Joos, Robert M. Key, Keith Lindsay, Ernst Maier-Reimer, Richard Matear, Patrick Monfray, Anne Mouchet, Raymond G. Najjar, Gian-Kasper Plattner, Keith B. Rodgers, Christopher L. Sabine, Jorge L. Sarmiento, Reiner Schlitzer, Richard D. Slater, Ian J. Totterdell, Marie-France Weirig, Yasuhiro Yamanaka, and Andrew Yool. 2005. Anthropogenic ocean acidification over the twenty-first century and its impact on calcifying organisms. *Nature* 437: 681–686.

5 Azevedo, Inês, Granger Morgan, Karen Palmer, and Lester Lave. 2010. The cost of reducing US residential energy use and CO_2 emissions, Department of Engineering and Public Policy, Carnegie Mellon University, 15pp.

6 Azevedo, Inês, M. Lima, M. Granger Morgan, and Fritz Morgan. 2009. The transition to solid-state lighting. *Proceedings of the IEEE* 97: 481–510.

7 National Research Council. 2010. *Real Prospects for Energy Efficiency in the United States: America's Energy Future Panel on Energy Efficiency Technologies.* Washington, DC: The National Academies Press.

8 Samaras, Constantine, and Kyle Meisterling. 2008. Life cycle assessment of greenhouse gas emissions from plug-in hybrid vehicles: Implications for policy. *Environmental Science & Technology* 42: 3170–3176.

9 Katzenstein, Warren, Emily Fertig, and Jay Apt. 2010. The variability of interconnected wind plants. *Energy Policy* 38: 4400–4410.

10 Katzenstein, Warren, and Jay Apt. 2009. Response to comment on "Air emissions due to wind and solar power." *Environmental Science & Technology* 43: 6108–6109.

11 Curtright, Aimee E., and Jay Apt. 2008. The character of power output from utility-scale photovoltaic systems. *Progress in Photovoltaics* 16: 241–247.

12 Curtright, Aimee E., M. Granger Morgan, and David Keith. 2008. Expert assessment of future photovoltaic technology. *Environmental Science & Technology* 42: 9031–9038.

13 See Food and Agriculture Organization of the United Nations. 2008. *The State of Food and Agriculture*. Rome: FAO; and Ajanovic, Amela. 2010. Biofuels versus food production: does biofuels production increase food prices? *Energy* 35(4): 2070–2076.

14 Searchinger, T., R. Heimlich, R. A. Houghton, F. Dong, A. Elobeid, J. Fabiosa, S. Tokgoz, D. Hayes, and T.-H. Yu. 2008. Use of US croplands for biofuels increases greenhouse gases through emissions from land-use change. *Science* 319: 1238–1240.

15 MIT. 2003, updated 2009. *The Future of Nuclear Power: An Interdisciplinary MIT Study*. http://web.mit.edu/nuclearpower (accessed December 26, 2010).

16 Rubin, E. S., R. N. Cooper, R. A. Frosch, T. H. Lee, G. Marland, A. H. Rosenfeld, and D. D. Stine. 1992. Realistic mitigation options for global warming. *Science* 257: 148–266.

17 See National Research Council of the National Academies. 2007. *Coal: Research and Development to Support National Energy Policy*. Washington, DC: The National Academies Press; and US Energy Information Agency. 2010. *Summary: U.S. Crude Oil, Natural Gas, and Natural Gas Liquids Proved Reserves, 2009*. http://www.eia.doe.gov/natural_gas/data_publications/crude_oil_natural_gas_reserves/cr.html (accessed December 26, 2010).

18 The initials CCS are used interchangeably to refer to "carbon capture and sequestration" or "carbon capture and storage" as the same activity is often termed, especially in Europe. The word *storage* is defined as "the act of storing" and the verb *to store* is defined as "v.t. to supply or stock with something, as for future use ... to accumulate or put away for future use." There is no way that more than a miniscule amount of the CO_2 injected deep underground in CCS will ever be removed for use in the future. Hence, in the interests of accuracy, we use the word *sequester* which is defined as "to remove or withdraw into solitude" (definitions from the Random House Unabridged Dictionary).

The choice of terminology is significant because it reflects, and creates, expectations about where carbon dioxide will actually reside in decades ahead. Compare, for example, the distinctions among "treatment, storage, and disposal" in the Federal Resource Conservation and Recovery Act (RCRA) and its implementing regulations, 42 U.S.C. s/s 321 et seq. (P.L. 94-580). In those cases, Congress and the US EPA distinguished "storage" from "disposal" because of a belief that the economic incentives to reclaim "stored" materials would allow a lower level of regulation than would be required in cases where there was no expectation that materials would be reclaimed; that is, in cases properly labeled as "disposal."

In the present case, and for similar reasons, we distinguish "storage" from "disposal." Thus, "storing" would be an appropriate term, if we expected to reclaim captured carbon, but "sequestration" is more appropriate if we want to focus on long-term exclusion from the atmosphere.

19 Svensson, Rickard, Mikael Odenberger, Filip Johnsson, and Lars Strömberg. 2004. Transportation systems for CO_2—Application to carbon capture and storage. *Energy Conversion and Management* 45: 2343–2353.

20 Bachu, Stefan. 2003. Screening and ranking of sedimentary basins for sequestration of CO_2 in geological media in response to climate change. *Environmental Geology* 44: 277–289.

2 TECHNOLOGY FOR CARBON CAPTURE AND GEOLOGIC SEQUESTRATION

1 Svensson, R., M. Odenberger, et al. 2004. Transportation systems for CO_2-application to carbon capture and storage. *Energy Conversion and Management* 45: 2343–2353.
2 Bachu, S. 2003. Screening and ranking of sedimentary basins for sequestration of CO_2 in geological media in response to climate change. *Environmental Geology* 44: 277–289.
3 See, for example, work by Zeman, F. 2008. Experimental results for capturing CO_2 from the atmosphere (R&D note). *American Institute of Chemical Engineers Journal* 54(5): 1396–1399; Mahmoudkhani, M., and D.W. Keith. 2009. Low-energy sodium hydroxide recovery for CO_2 capture from atmospheric air-thermodynamic analysis. *International Journal of Greenhouse Gas Control* 3(4): 376–384; and Stolaroff, J. K. 2006. Capturing CO_2 from ambient air: A feasibility assessment. Ph.D. dissertation, Carnegie Mellon University: 85.
4 Petroleum coke (petcoke) is a carbon rich residue that remains after upgrading of heavy oil.
5 Eide, L. I., and D.W. Bailey. 2005. Precombustion decarbonisation processes. *Oil & Gas Science and Technology-Revue de L' Institut Francais du Petrole* 60(3): 475–484.
6 A combined cycle gas turbine operates with both a topping cycle and a bottoming cycle. The topping cycle converts the potential energy in the hot combustion products into electricity via a generator. The bottoming cycle extracts thermal energy from the hot gas turbine exhaust by generating steam and converting the steam into electricity via steam turbine and generator.
7 Rubin, E. S., A. Marks, et al. 2010. *Carbon Capture: A Technology Assessment.* Washington, DC: Congressional Research Service.
8 Ibid.
9 The number of operating CO_2 capture pilot and commercial projects is rapidly changing. The current estimate is based on projects in the MIT CCS database (http://sequestration.mit.edu/tools/projects/index.html), the IEA RD&D database (http://www.ieaghg.org/index.php?/RDD-Database.html), and the US Department of Energy, National Energy Technology Laboratory CCS Database (http://www.netl.doe.gov/technologies/carbon_seq/database/index.html).
10 The number of plants is based on the Gasification Technologies Council Online Gasification Database (http://www.gasification.org/database1/search.aspx) as of February 8, 2011.
11 Global CCS Institute (GCCSI). 2011. *The Global Status of CCS: 2011.* Canberra, Australia: GCCSI.
12 See Baade, W. F., U. N. Parekh, et al. 2000. Hydrogen. *Kirk-Othmer Encyclopedia of Chemical Technology.* John Wiley & Sons; and Eide, L., I. and D. W. Bailey. 2005. Precombustion decarbonisation processes. *Oil & Gas Science and Technology-Revue De L Institut Francais Du Petrole* 60(3): 475–484.
13 See, for example, Rao, A. B. and E. S. Rubin. 2002. A technical, economic, and environmental assessment of amine-based CO_2 capture technology for power plant greenhouse gas control. *Environmental Science & Technology* 36(20): 4467–4475; and Idem, R. M. Wilson, et al. 2005. Pilot plant studies of the CO_2 Capture Performance of Aqueous MEA and Mixed MEA/MDEA Solvents at the University of Regina CO_2 Capture Technology Development Plant and the Boundary Dam CO_2 Capture Demonstration Plant. *Industrial and Engineering Chemistry Research* 45(8): 2414–2420.
14 Air is approximately 78% nitrogen, 21% oxygen, and the remainder is primarily argon and CO_2.
15 Thambimuthu, K., M. Soltanieh, et al. 2005. Capture of CO_2. In *IPCC Special Report on Carbon Dioxide Capture and Storage.* B. Metz, O. Davidson, H. de Coninck, M. Loos, and L. Meyer, eds. Cambridge: Cambridge University Press.

16 SO_x and NO_x in the flue gas result from oxidization of sulfur and nitrogen present in the fuel. NO_x levels are significantly lower than in traditional air-fired combustion because of the absence of N_2 in the air. See Thambimuthu, K., M. Soltanieh, et al. 2005. Capture of CO_2. In *IPCC Special Report on Carbon Dioxide Capture and Storage.* B. Metz, O. Davidson, H. de Coninck, M. Loos, and L. Meyer, eds. Cambridge: Cambridge University Press.

17 Anheden, M., J. Yan, et al. 2005. Denitrogenation (or oxyfuel concepts). *Oil & Gas Science and Technology - Rev. IFP* 60(3): 485–495.

18 Aspelund, A., and K. Jordal. 2007. Gas conditioning—The interface between CO_2 capture and transport. *International Journal of Greenhouse Gas Control* 1(3): 343–354.

19 Rubin, E. S., C. Chen, et al. 2007. Cost and performance of fossil fuel power plants with CO_2 capture and storage. *Energy Policy* 35(9): 4444–4454.

20 Formally, this is defined as $\eta_{ref}/\eta_{ccs} - 1$.

21 Koornneef, J., T. van Keulen, et al. 2008. Life cycle assessment of a pulverized coal power plant with post-combustion capture, transport and storage of CO_2. *International Journal of Greenhouse Gas Control* 2(4): 448–467.

22 Thambimuthu et al., Capture of CO_2.

23 Rubin, E. S., S. Yeh, et al. 2007. Use of experience curves to estimate the future cost of power plants with CO_2 capture. *International Journal of Greenhouse Gas Control* 1(2): 188–197.

24 Ibid.

25 Liu, H., and K. S. Gallagher. 2010. Catalyzing strategic transformation to a low-carbon economy: A CCS roadmap for China. *Energy Policy* 38(1): 59–74.

26 Morse, R. K., V. Rai, et al. 2009. *The Real Drivers of Carbon Capture and Storage In China and Implications for Climate Policy.* Working Paper #88, Stanford University: Program on Energy and Sustainable Development (PESD), Stanford University.

27 Rhodes, J. S., and D. W. Keith. 2005. Engineering economic analysis of biomass IGCC with carbon capture and storage. *Biomass and Bioenergy* 29(6): 440–450.

28 See, for example, Lackner, K. S., H.-J. Ziock, et al. 1999. Carbon dioxide extraction from air: Is it an option? *24th Annual Technical Conference on Coal Utilization and Fuel Systems.* Clearwater, FL; Baciocchi, R., G. Storti, et al. 2006. Process design and energy requirements for the capture of carbon dioxide from air. *Chemical Engineering and Processing* 45(12): 1047–1058; and Keith, D. 2009. Why capture CO_2 from the atmosphere. *Science* 325: 1654–1655.

29 Ibid.

30 See Searchinger, T., R. Heimlich, et al. 2008. Use of U.S. croplands for biofuels increases greenhouse gases through emissions from land-use change. *Science* 319(5867): 1238–1240; and Tilman, D., R. Socolow, et al. 2009. Beneficial biofuels—The food, energy, and environment trilemma. *Science* 325(5938): 270–271.

31 Keith, D. W., M. Ha-Duong, et al. 2006. Climate strategy with CO_2 capture from the air. *Climatic Change* 74(1–3): 17–45.

32 Ibid.

33 Zeman, F. S., and D. W. Keith. 2008. Carbon neutral hydrocarbons. *Philosophical Transactions of the Royal Society A-Mathematical, Physical & Engineering Sciences* 366(1882): 3901–3918.

34 See, for example, Baciocchi, R., G. Storti, et al. 2006. Process design and energy requirements for the capture of carbon dioxide from air. *Chemical Engineering and Processing* 45(12): 1047–1058; Keith, D. W., M. Ha-Duong, et al. 2006. Climate strategy with CO_2 capture from the air. *Climatic Change* 74(1–3): 17–45; and Zeman, F. 2007. Energy and material balance of CO_2 capture from ambient air. *Environmental Science & Technology* 41(21): 7558–7563.

35 Stolaroff, J. K., D. W. Keith, et al. 2008. Carbon dioxide capture from atmospheric air using sodium hydroxide spray. *Environmental Science & Technology* 42(8): 2728–2735.

36 Baciocchi, R., G. Storti, et al. 2006. Process design and energy requirements for the capture of carbon dioxide from air. *Chemical Engineering and Processing* 45(12): 1047–1058.

37 Keith et al., Climate strategy with CO_2 capture from the air.
38 See Baciocchi, R., G. Storti, et al. 2006. Process design and energy requirements for the capture of carbon dioxide from air. *Chemical Engineering and Processing* 45(12): 1047–1058; and Zeman, F. 2007. Energy and material balance of CO_2 capture from ambient air. *Environmental Science & Technology* 41(21): 7558–7563.
39 When reported energy requirements are all adjusted to include compression from atmospheric pressure to 15 MPa (150 bar), the lowest reported value is 13 GJ/t by Baciocchi, et al. and the highest is 18 GJ/t by Stolaroff. See Baciocchi, R., G. Storti, et al. 2006. Process design and energy requirements for the capture of carbon dioxide from air. *Chemical Engineering and Processing* 45(12): 1047–1058; and Stolaroff, J. K. 2006. Capturing CO_2 from ambient air: a feasibility assessment. Ph.D. dissertation, Carnegie Mellon University: 85.
40 Mahmoudkhani, M., and D. W. Keith. 2009. Low-energy sodium hydroxide recovery for CO_2 capture from atmospheric air—Thermodynamic analysis. *International Journal of Greenhouse Gas Control* 3(4): 376–384.
41 Belmabkhout, Y., R. Serna-Guerrero, et al. 2010. Amine-bearing mesoporous silica for CO_2 removal from dry and humid air. *Chemical Engineering Science* 65(11): 3695–3698.
42 Nikulshina, V., C. Gebald, et al. 2009. CO_2 capture from atmospheric air via consecutive CaO-carbonation and $CaCO_3$-calcination cycles in a fluidized-bed solar reactor. *Chemical Engineering Journal* 146(2): 244–248.
43 See, for example, Skovholt, O. 1993. CO_2 transportation system. *Energy Conversion and Management* 34(9–11): 1095–1103; and Svensson, R., M. Odenberger, et al. 2004. Transportation systems for CO_2-application to carbon capture and storage. *Energy Conversion and Management* 45: 2343–2353.
44 See, for example, Moritis, G. 2009. More CO_2-EOR projects likely as new CO_2 supply sources become available. *Oil and Gas Journal* 107(45): 41–47; and Moritis, G. 2010. CO_2 miscible, steam dominate enhanced oil. *Oil and Gas Journal* 108(14): 36–40.
45 Moritis, G. 2009. More CO_2-EOR projects likely as new CO_2 supply sources become available. *Oil and Gas Journal* 107(45): 41–47.
46 Recht, D. L. 1984. Design considerations for carbon-dioxide pipe lines. 1. *Pipe Line Industry* 61(3): 53–54.
47 Doctor, R., A. Palmer, et al. 2005. Transport of CO_2. In *IPCC Special Report on Carbon Dioxide Capture and Storage*. B. Metz, O. Davidson, H. de Coninck, M. Loos, and L. Meyer, eds. Cambridge: Cambridge University Press.
48 See, for example, McCoy, S. T., and E. S. Rubin. 2008. An engineering-economic model of pipeline transport of CO_2 with application to carbon capture and storage. *International Journal of Greenhouse Gas Control* 2(2): 219–229; and Chandel, M. K., L. F. Pratson, et al. 2010. Potential economies of scale in CO_2 transport through use of a trunk pipeline. *Energy Conversion and Management* 51(12): 2825–2834.
49 Newcomer, A., and J. Apt. 2008. Implications of generator siting for CO_2 pipeline infrastructure. *Energy Policy* 36(5): 1776–1787.
50 See, for example, Dahowski, R. T., and J. J. Dooley. 2004. Carbon management strategies for US electricity generation capacity: A vintage-based approach. *Energy* 29(9–10): 1589–1598; Dooley, J. J., R. T. Dahowski, et al. 2004. A CO_2-storage supply curve for North America and its implications for the deployment of carbon dioxide capture and storage systems. *7th International Conference on Greenhouse Gas Control Technologies*. Vancouver, Canada: Elsevier Science; Wise, M., J. Dooley, et al. 2007. Modeling the impacts of climate policy on the deployment of carbon dioxide capture and geologic storage across electric power regions in the United States. *International Journal of Greenhouse Gas Control* 1(2): 261–270; and van den Broek, M., A. Faaij, et al. 2008. Planning for an electricity sector with carbon capture and storage: Case of the Netherlands. *International Journal of Greenhouse Gas Control* 2(1): 105–129.
51 Dooley, J. J., R. T. Dahowski, et al. 2009. Comparing existing pipeline networks with the potential scale of future U.S. CO_2 pipeline networks. *Energy Procedia* 1(1): 1595–1602.

52 See Dahowski, R. T., and J. J. Dooley. 2004. Carbon management strategies for US electricity generation capacity: A vintage-based approach. *Energy* 29(9–10): 1589–1598; and ICF International. 2009. *Developing a Pipeline Infrastructure for CO_2 Capture and Storage: Issues and Challenges*. Report prepared for the Interstate Natural Gas Association of America, Washington, DC.

53 Parfomak, P. W., and P. Fogler. 2008. *Pipelines for Carbon Dioxide (CO_2) Control: Network Needs and Cost Uncertainties*. Washington, DC: Congressional Research Service.

54 See McCoy, S. T., and E. S. Rubin. 2008. An engineering-economic model of pipeline transport of CO_2 with application to carbon capture and storage. *International Journal of Greenhouse Gas Control* 2(2): 219–229; and Chandel, M. K., L. F. Pratson, et al. 2010. Potential economies of scale in CO_2 transport through use of a trunk pipeline. *Energy Conversion and Management* 51(12): 2825–2834.

55 Onshore, natural gas transmission system mileage accessed from http://phmsa.dot.gov/pipeline/library/data-stats on February 23, 2010; this estimate does not include natural gas distribution lines.

56 For further discussion, see House, K. Z., D. P. Schrag, et al. 2006. Permanent carbon dioxide storage in deep-sea sediments. *Proceedings of the National Academy of Sciences* 103(33): 12291–12295; Goldberg, D. S., T. Takahashi, et al. 2008. Carbon dioxide sequestration in deep-sea basalt. *Proceedings of the National Academy of Sciences* 105(29): 9920–9925; and Schrag, D. P. 2009. Storage of carbon dioxide in offshore sediments. *Science* 325(5948): 1658–1659.

57 National Energy Technology Lab. 2010. *Carbon Sequestration Atlas of the United States and Canada*. Pittsburgh, PA: US Department of Energy.

58 Ibid.

59 See Friedmann, S. J., J. J. Dooley, et al. 2006. The low cost of geological assessment for underground CO_2 storage: Policy and economic implications. *Energy Conversion and Management* 47: 1894–1901; Bachu, S., D. Bonijoly, et al. 2007. CO_2 storage capacity estimation: Methodology and gaps. *International Journal of Greenhouse Gas Control* 1(4): 430–443.

60 See Bachu, S., D. Bonijoly, et al. 2007. CO_2 storage capacity estimation: Methodology and gaps. *International Journal of Greenhouse Gas Control* 1(4): 430–443; and Bradshaw, J., S. Bachu, et al. 2007. CO_2 storage capacity estimation: Issues and development of standards. *International Journal of Greenhouse Gas Control* 1(1): 62–68.

61 See Birkholzer, J. T., and Q. L. Zhou. 2009. Basin-scale hydrogeologic impacts of CO_2 storage: Capacity and regulatory implications. *International Journal of Greenhouse Gas Control* 3(6): 745–756; and Person, M., A. Banerjee, et al. 2010. Assessment of basin-scale hydrologic impacts of CO_2 sequestration, Illinois basin. *International Journal of Greenhouse Gas Control* 4(5): 840–854.

62 For more details, see Craft, B., and M. Hawkins. 1959. *Applied Petroleum Reservoir Engineering*. Englewood Cliffs, NJ: Prentice-Hall; and Kearey, P., M. Brooks, et al. 2002. *An Introduction to Geophysical Exploration*. Oxford: Blackwell Publishing.

63 Most large projects will involve a large number of injection wells spread across the project footprint. Wells can be directionally drilled and may be horizontal. Use of directional drilling techniques can reduce the number of wells required and increase the fraction of the formation contacted by CO_2 as compared with the use of only traditional vertical wells.

64 Doughty, C., B. M. Freifeld, et al. 2008. Site characterization for CO_2 geologic storage and vice versa: the Frio brine pilot, Texas, USA as a case study. *Environmental Geology* 54(8): 1635–1656.

65 Benson, S., P. Cook, et al. 2005. Underground geological storage. In *IPCC Special Report on Carbon Dioxide Capture and Storage*. B. Metz, O. Davidson, H. de Coninck, M. Loos, and L. Meyer, eds. Cambridge: Cambridge University Press.

66 WRI. 2008. *Guidelines for Carbon Dioxide Capture, Transport, and Storage*. Washington, DC: World Resources Institute.

67 National Energy Technology Laboratory. 2010. *Site Screening, Selection, and Initial Characterization for Storage of CO_2 in Deep Geologic Formations*. Pittsburgh, PA: US Department of Energy.

68 DNV. 2009. *CO2QUALSTORE: Guideline for Selection and Qualification of Sites and Projects for Geological Storage of CO_2*. Hovik, Norway: Det Norske Veritas AS.

69 Apt, J., and B. Fischhoff. 2006. Power and people. *The Electricity Journal* 19(9): 17–25.

70 Byrd D. M., and C. R. Cothern. 2000. *Introduction to Risk Analysis: A Systematic Approach to Science-Based Decision Making*. Rockville, MD: Government Institutes. Chapter 11.

71 WRI, *Guidelines for Carbon Dioxide Capture*.

72 DNV, *CO2QUALSTORE*.

73 See, for example, a statistical analysis of the spatial and temporal distribution of abandoned wells penetrating the Viking Aquifer in the Alberta Basin in: Gasda, S. E., S. Bachu, et al. 2004. Spatial characterization of the location of potentially leaky wells penetrating a deep saline aquifer in a mature sedimentary basin. *Environmental Geology* 46(6–7): 707–720.

74 Benson et al., Underground geological storage.

75 Wilson, E. J., T. J. Johnson, et al. 2003. Regulating the ultimate sink: Managing the risks of geologic CO_2 storage. *Environmental Science & Technology* 37: 3476–3483.

76 WRI. 2010. *Guidelines for Community Engagement in Carbon Dioxide Capture, Transport, and Storage Projects*. Washington, DC: World Resources Institute.

77 See Benson, S., P. Cook, et al. 2005. Underground geological storage. In *IPCC Special Report on Carbon Dioxide Capture and Storage*. B. Metz, O. Davidson, H. de Coninck, M. Loos, and L. Meyer, eds. Cambridge: Cambridge University Press; and Bachu, S., D. Bonijoly, et al. 2007. CO_2 storage capacity estimation: Methodology and gaps. *International Journal of Greenhouse Gas Control* 1(4): 430–443.

78 See Birkholzer, J. T., Q. Zhou, et al. 2009. Large-scale impact of CO_2 storage in deep saline aquifers: A sensitivity study on pressure response in stratified systems. *International Journal of Greenhouse Gas Control* 3(2): 181–194; and Zhou, Q. L., J. T. Birkholzer, et al. 2010. Modeling basin- and plume-scale processes of CO_2 storage for full-scale deployment. *Ground Water* 48(4): 494–514.

79 Wilson, E. J., and D. Gerard, eds. 2007. *Carbon Capture and Sequestration: Integrating Technology, Monitoring and Regulation*. Oxford: Blackwell Publishing.

80 Benson, S. M., E. Gasperikova, et al. 2004. *Overview of Monitoring Techniques and Protocols for Geologic Storage Projects*. Stoke Orchard, UK: IEA Greenhouse Gas R&D Programme.

81 Nooner, S. L., O. Eiken, et al. 2007. Constraints on the in situ density of CO_2 within the Utsira formation from time-lapse seafloor gravity measurements. *International Journal of Greenhouse Gas Control* 1(2): 198–214.

82 Many studies have taken different approaches to estimate allowable leakage rates, including leakage rates that change with time: Teng, F., and D. Tondeur. 2007. Efficiency of carbon storage with leakage: Physical and economical approaches. *Energy* 32(4): 540–548; and van der Zwaan, B., and R. Gerlagh. 2009. Economics of geological CO_2 storage and leakage. *Climatic Change* 93(3–4): 285–309.

83 Benson, S. M. 2006. Monitoring carbon dioxide sequestration in deep geological formations for inventory verification and carbon credits. *SPE Annual Technical Conference and Exhibition*. San Antonio, TX: Society of Petroleum Engineers.

84 Ibid.

85 Cortis, A., C. M. Oldenburg, et al. 2008. The role of optimality in characterizing CO_2 seepage from geologic carbon sequestration sites. *International Journal of Greenhouse Gas Control* 2(4): 640–652.

86 DNV, *CO2QUALSTORE*.

87 See, for example, Chabora, E. R., and S. M. Benson. 2009. Brine displacement and leakage detection using pressure measurements in aquifers overlying CO_2 storage reservoirs. *Energy Procedia* 1(1): 2405–2412.

88 Benson, S. M., R. Hepple, et al. 2002. *Lessons Learned from Natural and Industrial Analogues for Storage*. Berkeley, CA: Lawrence Berkeley National Laboratory.

89 For an example, see Nayar, A. 2009. A lakeful of trouble. *Nature* 460(7253): 321–323.

90 Benson et al., *Lessons Learned*.

91 Bogen, K., E. A. Burton, et al. 2006. Source terms for CO_2 risk modeling and GIS/simulation based tools for risk characterization. *8th International Conference on Greenhouse Gas Control Technologies*. Trondheim, Norway: Elsevier Science.

92 Work that addresses basin scale issues includes: Birkholzer, J. T., and Q. L. Zhou. 2009. Basin-scale hydrogeologic impacts of CO_2 storage: Capacity and regulatory implications. *International Journal of Greenhouse Gas Control* 3(6): 745–756; and Person, M., A. Banerjee, et al. 2010. Assessment of basin-scale hydrologic impacts of CO_2 sequestration, Illinois basin. *International Journal of Greenhouse Gas Control* 4(5): 840–854.

93 The number of operating CO_2 sequestration projects is rapidly changing, and many projects are announced but their development status is unclear. The 150-project estimate is based on the US Department of Energy, National Energy Technology Laboratory (NETL) CCS Database (http://www.netl.doe.gov/technologies/carbon_seq/database/index.html). A similar database from the Global CCS Institute (GCCSI) CCS Project Database contains approximately half as many projects as the NETL database (http://www.globalccsinstitute.com/resources/data/status-ccs-project-database).

94 In the petroleum industry, acid gas refers to a mixture of H_2S and CO_2. Acid gas may be treated to recover sulfur, which can then be sold, but it is often more economical to reinject acid gas for disposal.

95 Bachu, S., and K. Haug. 2005. In-situ characteristics of acid-gas injection operations in the Alberta Basin, western Canada: Demonstration of CO_2 geological storage. In *Carbon Dioxide Storage in Deep Geologic Formations, Volume 2*, eds. S. Benson and D. Thomas, London: Elsevier.

96 Ibid.

97 Moritis, G. 2010. CO_2 miscible, steam dominate enhanced oil recovery processes. *Oil and Gas Journal* 108(14): 36–40.

98 Ibid.

99 Moritis, G. 2009. More CO_2-EOR projects likely as new CO_2 supply sources become available. *Oil and Gas Journal* 107(45): 41–47.

100 See Kovscek, A. R., and M. D. Cakici. 2005. Geologic storage of carbon dioxide and enhanced oil recovery. II. Cooptmization of storage and recovery. *Energy Conversion and Management* 46: 1941–1956; and Kovscek, A. R., and Y. Wang. 2005. Geologic storage of carbon dioxide and enhanced oil recovery. I. Uncertainty quantification employing a streamline based proxy for reservoir flow simulation. *Energy Conversion and Management* 46: 1920–1940.

101 There are 53 kg of CO_2 per thousand standard cubic feet (mscf) of CO_2; this corresponds to 3–8 mscf of CO_2 per barrel of oil. Estimates for net and gross CO_2 utilization are highly dependent on the project. See McCoy, S. T. 2008. The economics of CO_2 transport by pipeline and storage in saline aquifers and oil reservoirs. Ph.D. dissertation, Carnegie Mellon University: 267.

102 Moritis, More CO_2-EOR projects likely.

103 See Office of Technology Assessment. 1978. *Enhanced Oil Recovery Potential in the United States*. Washington, DC: Office of Technical Assessment; National Petroleum Council 1984. *Enhanced Oil Recovery*. Washington, DC: National Petroleum Council; Interstate Oil and Gas Compact Commission. 1993. *An Evaluation of Known Remaining Oil Resources in the United States: Project on Advanced Oil Recovery and the States*. Oklahoma City, OK: Interstate Oil & Gas Compact Comission; Kuuskraa, V., and R. Ferguson. 2008. *Storing CO_2 with Enhanced Oil Recovery*. Pittsburgh, PA: National Energy Technology Laboratory; and Kuuskraa, V., R. Ferguson, et al. 2009. *Storing CO_2 with*

Next Generation CO_2-EOR Technology. Pittsburgh, PA: National Energy Technology Laboratory.

104 See Shaw, J., and S. Bachu. 2002. Screening, evaluation, and ranking of oil reservoirs suitable for CO_2-flood EOR and carbon dioxide sequestration. *Journal of Canadian Petroleum Technology* 41(9): 51–61; and Sandrea, I., and R. Sandrea. 2007. Recovery factors leave vast target for EOR technologies, Pt. 2. *Oil and Gas Journal* 105(42): 39–42.

105 Advanced Resources International. 2010. *U.S. Oil Production Potential From Accelerated Deployment of Carbon Capture and Storage*. Arlington, VA: Advanced Resources International, Inc.

106 In 2003, the US Geological Survey estimated the inferred reserves (the difference between proved reserves and the remaining recoverable resources in known fields) to be added to proven reserves between 1991 and 2071 to be approximately 48 billion bbl oil in the lower-48 states, with a 90% confidence interval between 25 billion and 82 billion barrels oil. Between 1991 and 2009, revisions and extensions to proven reserves totaled 18 billion barrels in the lower-48 states based on EIA data. Thus, unless CO_2-EOR offers a step-change in recovery efficiency that is unprecedented in the history of oil production, it is statistically unlikely that 40 to 60 billion bbl would be added to reserves (i.e., less than a 50% percent probability) by use of EOR alone. See Attanasi, E. D., and T. C. Coburn. 2003. *Uncertainty and Inferred Reserve Estimates—The 1995 National Assessment*. Reston, VA: US Geological Survey.

107 Jaramillo, P., W. M. Griffin, et al. 2009. Life cycle inventory of CO_2 in an enhanced oil recovery system. *Environmental Science & Technology* 43(21): 8027–8032.

108 McCoy, S. T., M. Pollak, et al. 2010. Geologic sequestration through EOR: Policy and regulatory considerations for greenhouse gas accounting. *10th International Conference on Greenhouse Gas Control Technologies*. Amsterdam, Netherlands.

109 Pollak, M., S. J. Phillips, et al. 2011. Carbon capture and storage policy in the United States: A new coalition endeavors to change existing policy. *Global Environmental Change* 21: 313–323.

3 SITING CO_2 PIPELINES FOR GEOLOGIC SEQUESTRATION

1 Moritis, G. 2009. More CO_2-EOR projects likely as new CO_2 supply sources become available. *Oil and Gas Journal* 107(45): 41–47.

2 See Dahowski, R. T., and J. J. Dooley. 2004. Carbon management strategies for US electricity generation capacity: A vintage-based approach. *Energy* 29(9–10): 1589–1598; and ICF International. 2009. *Developing a Pipeline Infrastructure for CO_2 Capture and Storage: Issues and Challenges*. Washington, DC: Interstate Natural Gas Association of America.

3 See McCoy, S. T., and E. S. Rubin. 2008. An engineering-economic model of pipeline transport of CO_2 with application to carbon capture and storage. *International Journal of Greenhouse Gas Control* 2(2): 219–229; and Chandel, M. K., L. F. Pratson, et al. 2010. Potential economies of scale in CO_2 transport through use of a trunk pipeline. *Energy Conversion and Management* 51(12): 2825–2834.

4 Section 1(b) of the Natural Gas Act, 15 U.S.C. §717(b) (2006), defines the Act's scope:

> The provisions of this chapter shall apply to the transportation of natural gas in interstate commerce, to the sale in interstate commerce of natural gas for resale for ultimate public consumption for domestic, commercial, industrial, or any other use, and to natural-gas companies engaged in such transportation or sale . . . but shall not apply to any other transportation or sale of natural gas or to the local distribution

of natural gas or to the facilities used for such distribution or to the production or gathering of natural gas.

The FERC's core activities under the NGA include: (1) certification of jurisdictional pipeline and storage facilities (certification carries eminent domain authority); (2) regulation of rates, terms, and conditions for pipeline transportation and storage; and (3) oversight of wholesale sales for resale (although wholesale rates are largely deregulated).

5 15 U.S.C. § 717f(c)(1)(A).
6 15 U.S.C. § 717f(e).
7 Under 15 U.S.C. § 717f(h), if the holder and a property owner cannot agree to the terms of a right-of-way for the construction, operation, maintenance, or transportation of a natural gas pipeline, the holder "may acquire the same by the exercise of the right of eminent domain" in state or federal court.
8 15 U.S.C. § 717f(f).
9 15 U.S.C. § 717f(h).
10 See, for example, *Northern Natural Gas Co. v. Iowa Util. Bd.*, 377 F.3d 817 (8th Cir. 2004) (finding that federal regulations and the NGA occupy the field of extension, operation, and acquisition of natural gas facilities, thereby preempting any state authority to do so); see also *Schneidewind v. ANR Pipeline Company,* 485 U.S. 293 (1988) (the NGA preempts state attempts to regulate securities issued by interstate pipeline companies). A certificate does not have preemptive effect when a state is exercising federal delegated authority, such as that provided by the Clean Water Act. In such situations, the question is not one of preemption, but of which statute prevails.
11 *Cortez Pipeline Co.*, 7 FERC ¶ 61,024 (1979).
12 15 U.S.C. § 717(b).
13 49 U.S.C. § 60502.
14 49 U.S.C. §15301(a).
15 *Cortez Pipeline Co.*, 45 Fed. Reg. 85,177 at p. 85,178 (1980).
16 See *Exxon Corp. v. Lujan*, 970 F.2d 757 (10th Cir. 1992) (deferring to the BLM's authority to issue a ROW across federal lands under the MLA, thereby imposing common carrier obligations on pipeline operators).
17 30 U.S.C. § 185(a).
18 30 U.S.C.. at § 185(r)(2)(A).
19 7 FERC ¶ 61,024 (1979).
20 15 U.S.C. § 717a(5).
21 In the CCS context, it is unlikely that CO_2 will be mixed with any methane, so there is likely to be less of a question under the Natural Gas Act.
22 *Cortez Pipeline Co.*, 7 FERC ¶ 61,024, at p. 61,041.
23 Ibid.
24 *Cortez Pipeline Co.*, 7 FERC ¶ 61,024, at p. 61,042 (quoting *FPC v. Louisiana Power & Light Co.*, 406 U.S. 621 (1972)).
25 Ibid.
26 *Cortez Pipeline Co.—Petition for Declaratory Order—Commission Jurisdiction Over Transportation of Carbon Dioxide by Pipeline*, 45 Fed. Reg. 85,177 (1980). The ICC also ruled in the same order on a similar petition filed by the Atlantic Richfield Company, who sought, like the Cortez Pipeline Co., to transport CO_2 via pipeline from Colorado to Texas for tertiary recovery through EOR.
27 Ch 3591, 34 Stat. 584.
28 Ibid.
29 *Cortez Pipeline Co.*, 45 Fed. Reg. at 85178.
30 *Cortez Pipeline Co.*, 45 Fed. Reg. at 85178 (citing *Erlenbaugh v. U.S.*, 409 U.S. 239, 243–44 (1972)).
31 *Cortez Pipeline Co.*, 46 Fed. Reg. 18805 (1981).

32 49 U.S.C. 10501(a)(1)(C) (1976) as amended by Pub. L. No. 95–473, 92 Stat. 1359 (1978).

33 *Cortez Pipeline Co.*, 46 Fed. Reg. 18805. While the case indicated that the gas was transported as a supercritical fluid, the decision treats CO_2 as a gas at atmospheric pressure.

34 Pub. L. 104–88, 109 Stat. 803, 804 (1995).

35 U.S. General Accounting Office (GAO). 1998. *Issues Associated with Pipeline Regulation by the Surface Transportation Board.* Appendix 1, Pipelines and Commodities Under STB's Jurisdiction. GAO/RCED-98–99. Washington, DC: U.S. GAO.

36 See Vann, A. and P.W. Parfomak. 2008. *Regulation of Carbon Dioxide (CO_2) Sequestration Pipelines: Jurisdictional Issues.* RL 34307. Washington, DC: Congressional Research Service. 6 n.29. The authors suggest that Congress may need to enact specific legislation that provides for definitive CO_2 federal pipeline rate jurisdiction to prevent continued jurisdictional disclaimers.

37 Under US administrative law, an agency is free to change its interpretation of the statute it administers if there is a reasoned basis for its decision. See *Chevron USA, Inc. v. Natural Res. Def. Council*, 467 U.S. 837, 863 (1984); and *Motor Vehicles Mfg. Assoc. v. State Farm*, 463 U.S. 29, 42 (1983).

38 Compare 49 U.S.C. §15301(a) (2006) with prior 49 U.S.C. § 10501(a)(1)(C) (1978).

39 970 F.2d 757 (10th Cir. 1992).

40 43 U.S.C. § 1761.

41 43 U.S.C. § 1761(a)(2) (emphasis added).

42 43 U.S.C. at § 1761(b). An application for a ROW must contain information about the "effect on competition," which is considered in the terms and conditions of the grant of a ROW, but there are no specific non-discrimination or open access requirements.

43 *Chevron*, 467 U.S. 837.

44 *Exxon Corp. v. Lujan*, 970 F.2d 757, 761 (10th Cir. 1992).

45 49 U.S.C. § 15501(a).

46 49 U.S.C. § 15502.

47 49 U.S.C. § 15505.

48 49 U.S.C. § 15506.

49 49 U.S.C. § 15701; 49 C.F.R. §§ 1305.2–1305.3 (2009).

50 See, for example, 49 U.S.C. § 15501 (setting forth standards for pipeline rates, classifications, and rules for transportation or service provided by a "pipeline carrier").

51 49 U.S.C. § 15102(2).

52 See *U.S. v. Champlin Refining Co.*, 341 U.S. 290 (1951) (holding that the ICC could not regulate an oil company that transports its own products through its own pipeline, does not hold itself out as a public carrier, and does not transport products of any other company). See also *The Pipe Line Cases*, 234 U.S. 548, 562 (1914) (holding that the use of an oil pipeline for the sole purpose of moving oil across a state line from a company's own wells to its own refinery is not "transportation" within the meaning of the ICA).

53 Legislation introduced in the 111th Congress (2009–2010) would have extended the limits on STB's pipeline jurisdiction to exclude "when transporting a commodity other than water, oil, or natural or artificial gasses that are used primarily as a fuel or for other energy purposes." Surface Transportation Board Reauthorization Act of 2009, S. 2889, 111th Cong. § 502(a). The change would clarify that STB does not have jurisdiction over CO_2 pipelines used for CCS, which would be categorized as an "other energy purpose." Also, anthropogenic, or man-made, CO_2 may be categorized as an "artificial" gas. However, in § 502(b) the bill also directs the Comptroller General to submit a study to Congress "that analyzes the Federal and State regulatory framework to support the development of pipelines for the purposes of the capture, transportation, and sequestration of carbon dioxide." The provision faltered, but it could be reintroduced in subsequent Congresses.

54 49 U.S.C. § 15503(c).
55 U.S. GAO 1998.
56 15 U.S.C. § 717e(h).
57 This reluctance of allocating costs to entities not seen as beneficiaries is playing out in FERC policymaking efforts to engage in regional transmission planning and cost allocation. See, for example, *The Joint Comments of the North Carolina Public Utilities Commission, and the Public Staff of the North Carolina Public Utilities Commission*, Docket No. RM10–23 (filed Oct. 5, 2010); and *Comments of the Pennsylvania Public Utilities Commission*, Docket No. RM10–23 (filed Sept. 29, 2010).
58 16 U.S.C. § 824p(b)(1)(C).
59 *Piedmont Environmental Council v. FERC*, 558 F.3d 304 (4th Cir. 2009), *cert. denied*, 130 S.Ct. 1138 (2010).
60 Bliss, K., D. Eugene, et al. 2010. *A Policy, Legal, and Regulatory Evaluation of the Feasibility of a National Pipeline Infrastructure for the Transport and Storage of Carbon Dioxide*. Oklahoma, OK: Interstate Oil and Gas Compact Commission.
61 Williams, E., N. Greenglass, et al. 2007. *Carbon Capture, Pipeline and Storage: A Viable Option for North Carolina Utilities?* Durham, NC: Climate Change Policy Partnership, Duke University.
62 Bliss et al., *A Policy, Legal, and Regulatory Evaluation.*
63 Gresham, R. L., S.T. McCoy, et al. 2010. Implications of compensating property owners for geologic sequestration of CO_2. *Environmental Science & Technology* 44(8): 2897–2903.
64 49 U.S.C. § 60101 *et seq.* (2006).
65 49 U.S.C. § 60102(i) (2006).
66 49 U.S.C. § 195.2 (2009).
67 49 C.F.R. Part 195 (2011).
68 49 U.S.C. § 60104(c) (generally, states and local authorities "may not adopt or continue in force safety standards for interstate pipeline facilities or interstate pipeline transportation"); *Olympic Pipeline Co. v. City of Seattle*, 437 F.3d 872 (9th Cir. 2006) (finding that safety regulations imposed in addition to federal–state pipeline safety agreement were preempted by the Federal Pipeline Safety Act).
69 49 C.F.R. Part 195.
70 DOT Legislative Proposal. 2010. http://www.phmsa.dot.gov/staticfiles/PHMSA/DownloadableFiles/Pipeline%20Reauthorization%20Bill%209.15.2010.pdf (accessed March 31, 2012).
71 Ibid.
72 *Tex Nat. Res. Code Ann.* § 111.002 (Lexis 2010) ("A person is a common carrier subject to the provisions of this chapter if it: . . . (6) owns, operates, or manages, wholly or partially, pipelines for the transportation of carbon dioxide or hydrogen in whatever form to or for the public for hire, but only if such person files with the commission a written acceptance of the provisions of this chapter expressly agreeing that, in consideration of the rights acquired, it becomes a common carrier subject to the duties and obligations conferred or imposed by this chapter").
73 *Tex Nat. Res. Code Ann.* § 111.017.
74 *Tex Nat. Res. Code Ann.* § 111.014.
75 See 16 *Tex. Admin. Code* § 3.71 (LEXIS 2010).
76 *Tex Nat. Res. Code Ann* § 111.019.
77 *Tex Nat. Res. Code Ann* § 111.0194(a).
78 In § 111.002, the Texas statute lists seven categories of common carrier pipelines. Of those categories, only common carrier pipelines that transport coal require a certificate of public convenience (§§ 111.301–111.302). Common carrier pipelines that transport CO_2 do not require such a certificate.
79 The Texas common carrier statute is void of any discussion concerning the regulation of common carrier pipelines apart from coal pipelines. In FAQs issued by the Texas Railroad Commission (RRC), the RRC disclaims any authority to decide the route

a common carrier pipeline will take and asserts that the authority is vested with the pipeline's owner or operator. RRC Pipeline Eminent Domain and Condemnation FAQs are available at: http://www.rrc.state.tx.us/about/faqs/eminentdomain.php.

80 *N.M. Stat. Ann.* § 70–3-1 (Lexis 2010) ("The corporation commission [public regulation commission] may prescribe reasonable maximum rates for the transportation of oil and the products derived therefrom, where such products are transported by a pipeline common carrier from any point in New Mexico to an ultimate destination in New Mexico, provided, in the event the reasonableness of such rates are [is] contested in the manner provided by law, the burden of proof to show the unreasonableness of such rates shall be upon the person, firm, association or corporation contesting the same"). The New Mexico Constitution grants the New Mexico Public Regulation Commission the authority and responsibility to regulate "pipeline companies . . . in such manner as the legislature shall provide." N.M. Const. Art. XI, § 2 (Lexis 2010).

81 *NM Stat. Ann.* § 70–3-5(b). The statute states:

> Any person, firm, association or corporation may exercise the right of eminent domain to take and acquire the necessary right-of-way for the construction, maintenance and operation of pipelines, including microwave systems and structures and other necessary facilities for the purpose of conveyance of petroleum, natural gas, *carbon dioxide gas* and the products derived therefrom, but any such right-of-way shall in all cases be so located as to do the least damage to private or public property consistent with proper use and economical construction. Such land and right-of-way shall be acquired in the manner provided by the Eminent Domain Code [42A-1–1 NMSA 1978]. Pursuant to the requirements of Sections 42A-1–8 through 42A-1–12 NMSA 1978, the engineers, surveyors and other employees of such person, firm, association or corporation shall have the right to enter upon the lands and property of the state and of private persons and of private and public corporations for the purpose of making necessary surveys and examinations for selecting and locating suitable routes for such pipelines, microwave systems, structures and other necessary facilities, subject to responsibility for any damage done to such property in making surveys and examinations (§ 70–3–5(a)).

82 *NM Stat. Ann.* § 70–3–5(a).

83 *NM Stat. Ann.* § 42A-1–1–42A-1–33.

84 *NM Stat. Ann.* § 42A-1–22(C).

85 See 1983–1986 Op. Att'y Gen. N.M. 146 (1984) (discussing whether it was appropriate for the carbon dioxide pipelines to have eminent domain authority and finding that the legislature makes that determination. The petitioner raised the concern because CO_2 pipelines, when added to the New Mexico eminent domain power statute, were not used as a fuel by the general public, but for the extraction of oil and other petroleum products).

86 *Ohio Rev. Code Ann.* §§ 163.01–16.22 and 1723.01 (Lexis 2010).

87 *Ohio Rev. Code Ann.* § 1551.01(k).

88 *Ohio Rev. Code Ann.* § 4905.01 *et seq.*

89 15 *Pa. Cons. Stat.* § 1511 (Lexis 2010). PA general eminent domain power is found in 26 *Pa. Cons. Stat.* § 102 *et seq.*

90 *Ohio Rev. Code Ann.* § 1509.21 (amended by Senate Bill 165, effective June 30, 2010).

91 Kleinhenz & Associates. 2008. *The Role of CO_2 Enhanced Oil Recovery in Ohio's Economy and Energy Future.* http://www.pewclimate.org/docUploads/KleinhenCO2_Enhanced_Oil_Rec_rpt%208%2008_.pdf (accessed November 9, 2010).

92 National Energy Technology Lab. 2010. *Carbon Sequestration Atlas of the United States and Canada.* Pittsburgh, PA: US Department of Energy.

93 66 *Pa. Cons. Stat.* § 2815.

94 See Clinton Climate Initiative. 2009. *Viability of a Large-Scale Carbon Capture & Sequestration Network in Pennsylvania.* http://www.dcnr.state.pa.us/info/carbon/reports-research.aspx ("Pennsylvania Report") (accessed November 9, 2010). This was the third of three reports published in response to Act 129.
95 Ibid.
96 49 U.S.C. § 60105.
97 49 U.S.C. § 60106.
98 16 *Tex. Admin. Code* § 8.1 *et seq.* (Lexis 2010).
99 16 *Tex. Admin. Code* § 8.1(b)(2).
100 30 U.S.C. § 185(r).
101 The Californian Public Utilities Commission had delegated authority over safety in that instance.
102 Marston, P. M., and P. A. Moore. 2008. From EOR to CCS: The evolving legal and regulatory framework for carbon capture and storage. *Energy Law Journal* 29(2): 421–490.
103 15 U.S.C. § 717 *et seq.*
104 American Wind Energy Association (AWEA) and Solar Energy Industries Association (SEIA). 2009. *Green Transmission Superhighways: Building a Path to America's Clean Energy Future.* White paper. Washington, DC: AWEA and SEIA.
105 16 U.S.C. § 824p.
106 But see *Piedmont Envtl. Council v. FERC*, 558 F.3d at 308–09 (holding that FERC lacks the expansive permitting authority and control over state siting decisions that FERC had interpreted the statute to confer).
107 The type of risks that worry developers and investors include regulatory agency modifications of transportation contracts (as under section 5 of the Natural Gas Act), the imposition of open access transportations requirements. See also: *Pipeline Service Obligations and Revisions to Regulations Governing Self-Implementing Transportation Under Part 284 of the Commission's Regulations and Regulation of Natural Gas Pipelines After Partial Wellhead Decontrol*, Order 636, FERC Stats. & Regs., Regs. Preambles 1991–1996 ¶ 30.939 (1992), *order on reh'g*, Order No. 636-A, FERC Stats. & Regs., Regs. Preambles 1991–1996, ¶ 30,950 (1992), *order on reh'g*, Order No. 636-B, 61 FERC ¶ 61,272 (1992), *aff'd in part, rev'd in part sub nom., United Distribution Cos. v. FERC*, 88 F.3d 1105 (D.C. Cir. 1996), *cert. denied sub nom., Assoc. Gas Distribs. v. FERC*, 520 U.S. 1224 (1997), *order on remand*, Order No. 636-C, 78 FERC ¶ 61,186 (1997), *order on reh'g*, Order No. 636-D, 83 FERC ¶ 61,210 (1998); the Interstate Commerce Act section 3(1) (creating pro-rationing requirements for common carrier oil pipelines that prevent a pipeline from giving any one shipper undue preference. 49 U.S.C. app. § 3(1)).
108 Marston, From EOR to CCS.

4 PERMITTING GEOLOGIC SEQUESTRATION SITES

1 US EPA. 2010. Federal requirements under the Underground Injection Control (UIC) program for carbon dioxide (CO_2) geologic sequestration (GS) wells: Final rule. Fed. Reg. 75: 77230–303. December 10. Fed. Reg. Vol. 75, No. 237, Friday Dec. 10 2010: 2010; Vol. 40 C.F.R. Parts 124, 144, 145, 146, and 147, pp. 77230–77303.
2 42 U.S.C. § 300f *et seq.*
3 EPA defines a USDW as "An aquifer or portion of an aquifer that supplies any public water system or that contains a sufficient quantity of ground water to supply a public water system, and currently supplies drinking water for human consumption, or that contains fewer than 10,000 mg/l total dissolved solids and is not an exempted aquifer." See http://www.epa.gov/ogwdw/uic/glossary.html#u.

4 See EPA's definition of an exempted aquifer at: http://www.epa.gov/type/groundwater/uic/glossary.cfm#e.
5 EPA Underground Injection Control Program: Classes of Wells. http://www.epa.gov/safewater/uic/wells.html (accessed October 9, 2011).
6 Ibid.
7 Ibid.
8 Ibid.
9 Ibid.
10 Dougherty, C. C., and B. McLean. *Using Class V Experimental Technology Well Classification for Pilot Geologic Sequestration Projects—UIC Program Guidance (UICPG #83)*. Memo to EPA Water and Air Division Directors. Available at: http://www.epa.gov/ogwdw/uic/pdfs/guide_uic_carbonsequestration_final-0307.pdf (accessed October 8, 2011).
11 Kobelski, B. Company eyes precedent-setting permit under EPA CO_2 injection rules. *Inside EPA*. Posted August 16, 2011.
12 See http://www.epa.gov/ogwdw/uic/glossary.html#p.
13 Fed. Reg. Vol. 75, No. 237, Friday Dec. 10 2010: 2010; Vol. 40 C.F.R. Parts 124, 144, 145, 146, and 147, pp. 77230–77303.
14 40 C.F.R. 144.19.
15 40 C.F.R. 146.4.
16 While there is discussion of a "pressure front," in fact, in most formations, the pressure will decrease in a continuous manor, with no discrete "front."
17 Texas Railroad Commission. *Form H-1: Application to inject fluid into a reservoir productive of oil or gas*. http://www.rrc.state.tx.us/forms/forms/og/pdf/FormH-1p.pdf (accessed March 3, 2010).
18 US EPA. 2008. *Cost Analysis for the Proposed Federal Requirements Under the Underground Injection Control Program for Carbon Dioxide Geologic Sequestration Wells (Proposed GS Rule)*.
19 40 C.F.R. 146.90.
20 73 Fed. Reg. 43.495.
21 Note that there are a few wells in California where there is no USDW in the subsurface. In these cases, the UIC still regulates the injection.
22 40 C.F.R. 146.4.
23 Keith, D. W., J. A. Giardina, M. G. Morgan, et al. 2005. Regulating the underground injection of CO_2. *Environmental Science & Technology* 39: 499A–505A.
24 Pollak, M. F., and E. J. Wilson. 2009. Regulating geologic sequestration in the US: Early rules take divergent approaches. *Environmental Science & Technology* 43(9): 3035–3041.
25 Pollak, M. F., S. J. Phillips, and S. P. Vajjhala. 2011. Carbon capture and storage policy in the United States: A new coalition endeavors to change existing policy. *Global Environmental Change* 21: 313–323.
26 MT SB 498 (2009); ND SB 2095 (2009); WY HB 89 (2008).
27 WY HB 17 Sec. 1 (2010).
28 TX SB 1387 sec. 120.003(d) (2009).
29 KS HB 2419 sec. 3 (2007).
30 LA HB 661 sec. 1109(A).
31 ND HB 2095 sec. 38–22–17(6) (2009); MT SB 498 Sec. 4 (2009).
32 Dr. Lauren Fleischman contributed much of the text for this section.
33 See Huijts, N. M. A., C. J. H. Midden, and A. L. Meijnders. 2007. Social acceptance of carbon dioxide storage. *Energy Policy* 35(5): 2780–2789; Palmgren, C. R., M. G. Morgan, W. B. de Bruin, and D. W. Keith. 2004. Initial public perceptions of deep geological and oceanic disposal of carbon dioxide. *Environmental Science & Technology* 38(24): 6441–6450; Reiner, D., T. Curry, M. Figueiredo, et al. 2006. American exceptionalism? Similarities and differences in national attitudes toward energy policy and global warming. *Environmental Science & Technology* 40: 2093–2098; and Fleishman, L.A., W. B. de Bruin, and M. G. Morgan. 2010. Informed public preferences for electricity portfolios with CCS and other low-carbon technologies. *Risk Analysis* 30: 1399–1410.

34 WRI. 2010. *CCS and Community Engagement: Guidelines for Community Engagement in Carbon Dioxide Capture, Transport, and Storage Projects.* Washington, DC: World Resources Institute.
35 See WRI, *CCS and Community Engagement*; Foster, D. 2009. Victory for public safety in Carson City petcoke dispute. http://www.coalblock.org/coal-blog/17-blog-general/188-victory-for-public-safety-in-carson-city-petcoke-dispute.html (accessed September 20, 2009); and De Best-Waldhober, M., D. Daamen, and A. Faaij. 2009. Informed and uninformed public opinions of CO_2 capture and storage technologies in the Netherlands. *International Journal of Greenhouse Gas Control* 3(3): 322-332.
36 Slovic, P. 2000. *The Perception of Risk.* London, Sterling, VA: Earthscan Publications. xxxvii.
37 See Bord, R. J., and R. E. O'Connor. 1990. Risk communication, knowledge, and attitudes: Explaining reactions to a technology perceived as risky. *Risk Analysis* 10(4): 499–506; Maharik, M., and B. Fischhoff. 1993. Risk knowledge and risk attitudes regarding nuclear energy sources in space. *Risk Analysis* 13(3): 345–353; and Allum, N., P. Sturgis, D. Tabourazi, and I. Brunton-Smith. 2008. Science knowledge and attitudes across cultures: A meta-analysis. *Public Understanding of Science* 17(1): 35–54.
38 Fleischman, L. A., W. B. de Bruin, and M. G. Morgan. 2010. Informed public preferences for electricity portfolios with CCS and other low-carbon technologies. *Risk Analysis* 30: 1399–1410.
39 See De Best-Waldhober, M., D. Daamen, and A. Faaij. 2009. Informed and uninformed public opinions of CO_2 capture and storage technologies in the Netherlands. *International Journal of Greenhouse Gas Control* 3(3): 322–332; Wallquist, L., V. H. M. Visschers, and M. Siegrist. 2009. Lay concepts on CCS deployment in Switzerland based on qualitative interviews. *International Journal of Greenhouse Gas Control* 3(5): 652–657; and Siegrist, M., H. Gutscher, and T. C. Earle. 2005. Perception of risk: The influence of general trust, and general confidence. *Journal of Risk Research* 8(2): 145–156.
40 Fed. Reg. Vol. 75, No. 237. December 10, 2010: 77273.
41 40 C.F.R. 25 and 124.
42 40 C.F.R. 124.10.
43 Morgan, M. G., B. Fischhoff, A. Bostrom, and C. Atman. 2002. *Risk Communication: A Mental Models Approach.* New York, NY: Cambridge University Press.
44 WRI, *CCS and Community Engagement.*
45 Wade, S., and S. Greenberg. 2009. Afraid to start because the outcome is uncertain?: Social site characterization as a tool for informing public engagement efforts. *Energy Procedia* 1(1): 4641–4647.

5 LEARNING FROM AND ADAPTING TO CHANGES IN GEOLOGIC SEQUESTRATION TECHNOLOGY

1 Coglianese, C., J. Nash, and T. Olmstead. 2004. Performance-based regulation: Prospects and limitations in health, safety, and environmental protection. *Administrative Law Review* 55: 705–728.
2 May, P. J. 2003. Performance-based regulation and regulatory regimes: The saga of leaky buildings. *Law & Policy* 25(4): 381–401.
3 Ibid.
4 National Performance Review. 1995. *Reinventing Environmental Regulation.* http://govinfo.library.unt.edu/npr/library/rsreport/251a.html (accessed January 25, 2011).
5 May, P. J. 2003. Performance-based regulation and regulatory regimes: The saga of leaky buildings. *Law & Policy* 25(4): 381–401.
6 Ibid.
7 Ibid.

8 Sappington, D. E. M., J. P. Pfeifenberger, P. Hanser, and G. N. Basheda. 2001. The state of performance-based regulation in the U.S. electric utility industry. *The Electricity Journal* 14(8): 71–79.
9 Navarro, P. 1996. The simple analytics of performance-based ratemaking: A guide for the PBR regulator. *The Yale Journal on Regulation* 13(105): 105–161.
10 May, Performance-based regulation and regulatory regimes.
11 Laxton, A. 2010. Govt stumped as leaky home bill skyrockets. *NZHerald* online edition, February 27. http://www.nzherald.co.nz/nz/news/article.cfm?c_id=1&objectid=10628835 (accessed January 4, 2011).
12 New Zealand Building Act. 1991. (1990 No. 150) Part 2 Sec. 6.
13 May, Performance-based regulation and regulatory regimes.
14 Ibid.
15 Ibid.
16 McCray, L. E., K. A. Oye, and A. C. Petersen. 2010. Planned adaptation in risk regulation: An initial survey of US environmental, health, and safety regulation. *Technological Forecasting and Social Change* 77(6): 951–959.
17 Ibid.
18 Ibid.
19 Blais, L. E., and W. E. Wagner. 2008. Emerging science, adaptive regulations, and the problem of rulemaking ruts. *Texas Law Review* 86: 1701–1739.
20 McCray, Planned adaptation in risk regulation.
21 Ibid.
22 21 U.S.C. § 355(k); 21 C.F.R. §§ 314.80-.81.
23 21 C.F.R. § 314.150(a)(2).
24 Furberg, C. D., A. A. Levin, P. A. Gross, R. S. Shapiro, and B. L. Strom. 2006. The FDA and drug safety: A proposal for sweeping changes. *Archives of Internal Medicine* 166(18): 1938–1942.
25 42 U.S.C. § 7408(a)(1)(A).
26 42 U.S.C. § 7409.
27 42 U.S.C. § 7409(d)(1).
28 42 U.S.C. § 7409(d)(2). Details on the CASAC can be found at http://yosemite.epa.gov/sab/sabpeople.nsf/WebCommittees/CASAC.
29 McCray, L. E., and K. A. Oye. 2007. Adaptation and anticipation: Learning from policy experience. Working paper presented at the TransAtlantic Uncertainty Colloquium, University of Georgia, October 10–12, 2007. http://www.modeling.uga.edu/tauc/background_material/MCCRAYOYE-OCT6.pdf.
30 McCray, Planned adaptation in risk regulation.
31 Costa, D., et al. 2004. *Particulate Matter Research Program: Five Years of Progress*. EPA 600/R-04/058. http://www.epa.gov/airscience/pdf/pm_research_program_five_years_of_progress.pdf (accessed January 4, 2011).
32 McCray, Planned adaptation in risk regulation.
33 See US EPA. 2010. Federal requirements under the Underground Injection Control (UIC) program for carbon dioxide (CO_2) geologic sequestration (GS) wells: Final rule. Fed. Reg. 75: 77230–303. December 10; and US EPA. 2010. Mandatory reporting of greenhouse gases: Injection and geologic sequestration of carbon dioxide: Final rule. Fed. Reg. 75: 75060–89. December 1.
34 40 C.F.R. §§ 146.83–4.
35 40 C.F.R. § 146.83(a).
36 40 C.F.R. §§ 146.86 and 146.88.
37 40 C.F.R. § 146.86(b)(3).
38 40 C.F.R. § 146.90.
39 40 C.F.R. § 98.448(a)(3).
40 US EPA, Mandatory reporting of greenhouse gases.
41 US EPA Office of Air and Radiation. 2010. General technical support document

for injection and geologic sequestration of carbon dioxide: Subparts RR and UU. Greenhouse gas reporting program.

42 40 C.F.R. § 146.84(e).

43 40 C.F.R. § 146.90(j).

44 US EPA, Federal requirements under the Underground Injection Control (UIC) program.

45 McCray et al., Planned adaptation in risk regulation.

6 ACCESS TO PORE SPACE FOR GEOLOGIC SEQUESTRATION

1 Portions of this chapter are based on CCSReg Project. 2009. Chapter 5: Access to and use of pore space for CCS in the deep subsurface. *Carbon Capture and Sequestration: Framing the Issues for Regulation: An Interim Report from the CCSReg Project*. Pittsburgh, PA: Department of Engineering and Public Policy, Carnegie Mellon University, http://www.ccsreg.org/pdf/ccsreg_3_9.pdf (accessed October 9, 2011); and Anderson, Owen L. and R. Lee Gresham. 2011. *Legal and Commercial Models for Pore Space Access and Use for Geologic CO_2 Sequestration*. University of Pittsburgh Law Review, 72.

2 See Anderson, Owen L. 2009. Geologic CO_2 sequestration: Who owns the pore space? 9 *Wyo. L. Rev.* 9: 97; Anderson, Owen L. 2011. Lord Coke, the Restatement, and modern subsurface trespass law. *Texas Journal of Oil, Gas and Energy Law* 6.2; and Anderson, Owen L., and R. Lee Gresham. 2011. Legal and commercial models for pore space access and use for geologic CO_2 sequestration. *University of Pittsburgh Law Review* 72.

3 US Department of the Interior (DOI). 2009. *Report to Congress: Framework for Geological Carbon Sequestration on Public Land.*

4 Reese, A. *Climate: States Moving to Clarify Landowners' Rights over CO_2 Storage Space*. Land letter, February 26, 2009. http://www.eenews.net (accessed July 21, 2010) (noting that some states are also attempting to write clauses in GS legislation that protect existing resources and property interests).

5 US Energy Information Administration (EIA). 2010. *Average Depths of Crude Oil and Natural Gas Wells*. http://tonto.eia.doe.gov/dnav/pet/pet_crd_welldep_s1_a.htm (accessed July 21, 2010).

6 Doughty, C., and K. Pruess. 2004. Modeling supercritical carbon dioxide injection in heterogeneous porous media. *Vadose Zone Journal* 3: 837–847.

7 Benson, S., P. Cook, et al. 2005. Underground geological storage. In *IPCC Special Report on Carbon Dioxide Capture and Storage*. B. Metz, O. Davidson, H. de Coninck, M. Loos, and L. Meyer, eds. Cambridge: Cambridge University Press (stating that the presence of CO_2 in the basin can lead to corrosion problems and can change the composition such that plugging, erosion, and processing problems arise).

8 Doughty and Pruess, Modeling supercritical carbon dioxide injection.

9 See Benson, S. M., and D. R. Cole. 2008. CO_2 sequestration in deep sedimentary formations. *Elements* 4(5): 325–331; and Gunter, W. D., S. Bachu, and S. Benson. 2004. The role of hydrogeological and geothermal trapping in sedimentary basins for secure storage of carbon dioxide. In *Geological Storage of Carbon Dioxide*. S. J. Baines, and R. H. Worden, eds. Trowbridge, UK: Cromwell Press, 129–244.

10 WRI (World Resources Institute). 2009. *CCS Guidelines: Guidelines for Carbon Dioxide Capture, Transport, and Storage*. Washington, DC: WRI Press.

11 See Veil, J. A., M. G. Puder, D. Elcock, and R. J. Redweik. 2004. *A White Paper Describing Produced Water from Production of Crude Oil, Natural Gas, and Coal Bed Methane*. Argonne, IL: Argonne National Laboratory; and Puder, M. G., and J. A. Veil. 2006. *Offsite Commercial Disposal of Oil and Gas Exploration and Production Waste: Availability, Options, and Costs.* Argonne, IL: Argonne National Laboratory.

12 Doughty and Pruess, Modeling supercritical carbon dioxide injection.
13 WRI, *CCS Guidelines*.
14 Ibid.
15 Ibid.
16 Klass, A. B., and E. J. Wilson. 2010. Climate change, carbon sequestration, and property rights. *2010 U. Ill. L. Rev* 363.
17 Doughty and Pruess, Modeling supercritical carbon dioxide injection.
18 US Energy Information Administration (EIA). 2010. *Weekly Natural Gas Storage Report*. http://www.eia.doe.gov/oil_gas/natural_gas/ngs/ngs.html (accessed July 21, 2010).
19 Klass and Wilson, Climate change, carbon sequestration, and property rights.
20 Bachu, S. 2000. Sequestration of CO_2 in geologic media: Criteria and approach for site selection in response to climate change. *Energy Conversion and Management* 41(9): 953–970.
21 WRI, *CCS Guidelines*.
22 US Environmental Protection Agency (EPA). 2001. *Class I Underground Injection Control Program: Study of Risks Associated with Class I Underground Injection Wells*. Washington, DC: US EPA.
23 US Environmental Protection Agency (EPA). *Industrial & Municipal Waste Disposal Wells (Class I)*. http://www.epa.gov/safewater/uic/wells_class1.html (accessed July 21, 2010).
24 See US Environmental Protection Agency, *Class I Underground Injection Control Program*; and Keith, D.W., et al. 2005. Regulating the underground injection of CO_2. *Environmental Science & Technology* 39(24): 499A–505A.
25 40 C.F.R §§ 144–146.
26 US Environmental Protection Agency (EPA). 2010. *Federal Requirements Under the Underground Injection Control (UIC) Program for Carbon Dioxide (CO_2) Geologic Sequestration (GS) Wells: Final Rule*. Fed. Reg. 75: 77230–303. December 10.
27 40 C.F.R §§ 144–146.
28 Ibid.
29 See Coke, *Institutes of the Lawes of England*, Ch. 1, § 1(4a) (19th ed. 1832); Blackstone, *Commentaries on the Laws of England* 18 (Lewis ed. 1902); and Kent, *Commentaries on American Law* 621 (Gould ed. 1896).
30 Means the mineral and surface interests are held by a single owner. Garner, B. A. 2004. *Black's Law Dictionary*. 8th ed. St. Paul, MN: Thomson West.
31 See, for example, *Sunray Oil v. Cortez Oil Co.*, 112 P.2d 792 (Okla. 1941) (surface owner has the right to grant permission to inject wastewater into the subsurface as long as there is not interference with the mineral estate's recovery of oil and gas); *Central Kentucky Natural Gas Co. v. Smallwood*, 252 S.W.2d 866 (Ky. Ct. App. 1952) (the mineral owner has the authority to grant a gas-storage lease); *Tate v. United Fuel Gas Co.*, 71 S.E.2d 65 (W. Va. 1952) (based on exception language in the deed, the oil and gas owner did not own "clay, sand, or stone," and thus did not own the spaces in the storage formation, which were devoid of recoverable gas); *Emeny v. United States*, 412 F.2d 1319, 1324 (Ct. Cl. 1969) (ownership of "all gas in or under, or which may hereafter be found in or under and produced from the lands" did not include the right to store helium produced from other lands; such right remained with the non-mineral owner); *Miles v. Home Gas Co.*, 316 N.Y. Supp. 2d 908 (N.Y. Sup. Ct. App. Div. 1970) (surface owner); *Humble Oil & Refining Co. v. West*, 508 S.W.2d 812, 815 (Tex. 1974) (surface owner, citing *Emeny v. United States*); *Ellis v. Arkansas Louisiana Gas Co.*, 450 F. Supp. 412 (E.D. Okla. 1978) (storage company must obtain permission from the surface owner to store natural gas); *United States v. 43.42 Acres of Land*, 520 F. Supp. 1042, 1043, 1046 (W.D. La. 1981) (stated "the mineral owner cannot be considered to have ownership of the subsurface strata containing the spaces where the minerals are found"); *Rook v. James E. Russell Petroleum, Inc.*, 679 P.2d 158, 166–167 (Kan. 1984) (although Kansas has not directly addressed pore-space ownership, such rights are considered severable from the right to produce oil and gas); *Miss. River Transmission Corp. v. Tabor*, 757 F.2d 662, 672 (5th Cir. 1985)

(surface owner owns storage rights); *Grynberg v. City of Northglenn*, 739 P.2d 230 (Colo. 1987) (court's dubious holding that surface owner did not have a right to test for coal to determine suitability of land for a surface water reservoir suggests the possibility that mineral owner controls access to pore spaces); *Int'l Salt Co. v. Geostow*, 697 F. Supp. 1258 (W.D.N.Y. 1988) (salt miner given continued access to mined salt cavern to facilitate ongoing salt mining operation possibly suggesting that mineral owner controls access to pore spaces); *Mapco, Inc. v. Carter*, 808 S.W.2d 262 (Tex. App. 1991) (mineral owner, but inconsistent with *West*, above); *Chance v. BP Chemicals, Inc.*, 670 N.E.2d 985 (Ohio 1996) (the court found that "ownership rights in today's world are not as clear-cut as they were before the advent of airplanes and injections wells"); *Dep't of Transp. v. Goike*, 560 N.W.2d 365, 365 (W.D. Mich. 2006) (court held that "the storage space, once it has been evacuated of the minerals and gas, belongs to the surface owner); Stamm, Alan. 1957. Legal problems in the underground storage of natural gas. 36 Tex. L. Rev. 161 (1957); Scott, R. R. 1966. Underground storage of natural gas: A study of legal problems. 19 Okla. L. Rev. 47 (1966); McGaha, F. 1986. Underground gas storage: Opposing rights and interests. 46 La. L. Rev. 871 (1986); Wilson, E. J., and M. A. de Figueiredo. 2006. Geologic carbon dioxide sequestration: An analysis of subsurface property law. 36 ELR 10114 (2006); as well as Anderson 2009, Klass and Wilson 2010, Gresham and Anderson 2011, all cited above.

32 Recent white papers and law review articles have analyzed whether, in the first instance, the surface owner or the mineral owner on split-estate land has property rights in the pore space. While most of these papers and articles conclude that the surface owner would prevail over the mineral owner in most cases, the issue is far from resolved. Cooney, D. 2005. *Analysis of Property Rights Issues Related to Underground Storage Space Used for Geologic Sequestration of Carbon Dioxide.* IOGCC Task Force on Carbon Capture and Geologic Storage; as well as Wilson and de Figueiredo, Geologic carbon dioxide sequestration; Anderson, Geologic CO_2 sequestration; Klass and Wilson, Climate change, carbon sequestration, and property rights; Anderson and Gresham, *Legal and Commercial Models*.

33 de Figueiredo, M. A. 2007. The liability of carbon dioxide storage. Ph.D. dissertation, MIT Engineering Systems Division; as well as Wilson and de Figueiredo, Geologic carbon dioxide sequestration.

34 See Anderson, Geologic CO_2 sequestration, 97; and Duncan, I., S. Anderson, and J. Nicot. 2009. Pore space ownership issues for CO_2 sequestration in the U.S. *Energy Procedia* 1: 4427–4431.

35 de Figueiredo, The liability of carbon dioxide storage.

36 S.B. 498, 61st Leg., Reg. Sess. (Mont. 2009) (codified at Mont. Code Ann. § 82–11–180 (2009)); S.B. 2319, 61st Leg. Assem., Reg. Sess. (N.D. 2009) (codified at N.D. Cent. Code § 47–31–02 (2006 & Supp. 2009)); H.B. 89, 59th Leg., Budget Sess. (Wyo. 2008) (codified at Wyo. Stat. Ann. § 34–1–152 (2009)).

37 Wyo. H.B. 89 (codified at Wyo. Stat. Ann. § 34–1–452(a); H.R. 57, 60th Leg., Gen. Sess. (Wyo. 2009).

38 Wyo. Stat. Ann. § 34–1–452(a); H.R. 57, 60th Leg., Gen. Sess. (Wyo. 2009).

39 N.D. S.B. 2319 (codified at N.D. Cent. Code § 47–31–04).

40 Wyo. Stat. Ann. § 34–1–452(a); H.R. 57, 60th Leg., Gen. Sess. (Wyo. 2009).

41 S.B. 498, 61st Leg., Reg. Sess. (Mont. 2009) (enacted at Mont. Code Ann. § 82–11–180 (1009)).

42 H.B. 2860, 79th Leg., 1st Sess. (W. Va. 2009) (enacted in pertinent part at W. Va. Code 22–11A–6 (2009)).

43 US Constitution Amendment V.

44 *Armstrong v. United States*, 364 U.S. 40 (1960).

45 *Philips v. Wash. Legal Found.*, 524 U.S. 156 (1998).

46 *Lucas v. S.C. Coastal Council*, 505 U.S. 1003 (1992) (looking to "background principles" of the state's law of property and nuisance as an exception to the per se takings rule for

regulations that deprive a landowner of all economic use of property); *Klamath Irrigation Dist. v. United States*, 67 Fed. Cl. 504 (Fed. Cl. 2005).

47 *Bair v. United States*, 515 F.3d 1323 (Fed. Cir. 2008) (recognizing that "state-created property interests may be limited by federal laws" and that federal law can constitute "background principles" that can prevent a per se takings claim).

48 *Lingle v. Chevron U.S.A. Inc.*, 544 U.S. 528 (2005).

49 See *Lingle v. Chevron U.S.A. Inc.*, 544 U.S. 528 (2005); and *Loretto v. Teleprompter Manhattan CATV Corp.*, 458 U.S. 419 (1982).

50 See *Loretto v. Teleprompter Manhattan CATV Corp.*, 458 U.S. 419 (1982); and *Lucas v. S.C. Coastal Council*, 505 U.S. 1003 (1992).

51 *Tahoe-Sierra Pres. Council, Inc. v. Tahoe Reg'l Planning Agency*, 535 U.S. 302 (2002).

52 *Bair v. United States*, 515 F.3d 1323 (Fed. Cir. 2008).

53 *Lucas v. S.C. Coastal Council*, 505 U.S. 1003 (1992).

54 *Penn Central Transportation Co. v. New York City*, 438 U.S. 104 (1978).

55 *Lucas v. S.C. Coastal Council*, 505 U.S. 1003 (1992).

56 *Brown v. Legal Found. of Wash.*, 538 U.S. 216 (2003).

57 *Penn Central Transportation Co. v. New York City*, 438 U.S. 104 (1978).

58 *Causby*, 328 U.S. 256 (1946).

59 *Brown v. Legal Found. of Wash.*, 538 U.S. 216 (2003).

60 Ibid.

61 See Klass and Wilson, Climate change, carbon sequestration, and property rights; Anderson, O. L. 2010. Subsurface "trespass": A man's subsurface is not his castle; 49 Washburn L.J. 247 (2010); and Anderson and Gresham, *Legal and Commercial Models*.

62 *Loretto v. Teleprompter Manhattan CATV Corp.*, 458 U.S. 419 (1982).

63 See note 62

64 See note 62.

65 See note 62.

66 *Loretto v. Teleprompter Manhattan CATV Corp.*, 446 N.E.2d 428 (1983).

67 See Wilson and de Figueiredo, Geologic carbon dioxide sequestration; Anderson, Geologic CO_2 sequestration; Klass and Wilson, Climate change, carbon sequestration, and property rights; Anderson, Subsurface "trespass"; and Anderson and Gresham, *Legal and Commercial Models*.

68 *Columbia Gas v. Exclusive Natural Gas Storage Easement*, 747 F. Supp. 401 (N.D. Ohio 1990).

69 *Pac. Gas & Electric Co. v. Zuckerman*, 234 Cal. Rptr. 630 (Ct. App. 1987); *Iroquois Gas Corp. v. Gernatt*, 281 N.Y.S.2d 896 (N.Y. App. Div. 1967).

70 Colo. Rev. Stat. Ann. § 34–64–104 (condemnation is not allowed unless the formation is non-productive of gas in commercial quantities under either primary or secondary recovery methods); 220 Ill. Comp. Stat. 15/2 (the same as Colorado); Mo. Rev. Stat. § 393.460 (allowing condemnation of a formation for gas storage purposes unless the formation contains natural gas that may be produced in commercial quantities); W. Va. Code § 54–1–2(a)(3) (condemnation of underground storage facilities is permitted when previous exploration has shown the formation has ceased to produce, or has been proved to be non-productive of, gas in substantial quantities); Ark. Code Ann. § 15–72–604(a)(1) (no condemnation is permitted if the formation is producing, or is capable of producing, in paying quantities through any known recovery method); Mont. Code Ann. § 82–10–303(1)(b) (gas-bearing formations are not subject to condemnation unless recoverable volumes of native gas have been produced, or the formation has greater value or utility as storage than for production of relatively small amounts of remaining native gas compared with original volumes); Okla. Stat. Ann. tit. 52 § 36.3(b) (no condemnation of a formation is permitted unless the volume of native gas originally in place is substantially depleted, and the formation has greater utility as a storage facility than for production of relatively small volumes of remaining

gas); Neb. Rev. Stat. § 57–605 (if any recoverable native gas remains, the use of a formation for storage must be determined to be in the public interest because the formation has greater value as storage facility than for production of remaining volumes of native gas).

71 *Anderson v. Beech Aircraft*, 699 P.2d 1023 (Kan. 1985).
72 Cal. Pub. Res. Code § 3643(b) (allowing forced unitization if at least three-fourths of royalty interests and working interests agree to unitize).
73 16 Tex. Admin. Code § 3.40.
74 *Railroad Commission v. Manziel*, 361 S.W.2d 560 (Tex. 1962).
75 16 Tex. Admin. Code § 3.40.
76 Ibid.
77 Ibid.
78 See Wilson and de Figueiredo, Geologic carbon dioxide sequestration; Klass and Wilson, Climate change, carbon sequestration, and property rights; Anderson, Subsurface "trespass"; and Anderson and Gresham, *Legal and Commercial Models*.
79 See note 78.
80 See note 78.
81 *Chance v. BP Chemicals, Inc.*, 670 N.E.2d 985 (Ohio 1996).
82 See note 78.
83 See note 78.
84 See note 78.
85 See note 78.
86 Kiel, P. J., and G. A. Thomas. 2003. Banking groundwater in California: Who owns the aquifer storage space? *Nat. Resources and Environment* 18: 25.
87 Taguchi, T. L. 2003. Whose space is it anyway? Protecting the public interest in allocating storage space in California's groundwater basins. 32 Sw. U. L. REV. 117 (2003).
88 *Alameda County Water Dist. v. Niles Sand & Gravel Co.*, 112 Cal. Rptr. 846 (Ct. App. 1974); *Board of County Commissioners v. Park County Sportsmen's Ranch*, LLP, 45 P.3d 693 (Colo. 2002).
89 US Constitution Amendment V.
90 See Gresham, R. L., McCoy, S. T., Apt, J., and Morgan, M. G. 2010. Implications of compensating property owners for geologic CO_2 sequestration. *Environ. Sci. Technol.* 44(8): 2789–2903; and Gresham, R. Lee. 2010. Chapter 4 in *Geologic CO_2 Sequestration and Subsurface Property Rights: A Legal and Economic Analysis*, doctoral thesis, Department of Engineering and Public Policy, Carnegie Mellon.
91 US Constitution Amendment V.
92 McCoy, S. T., and E. S. Rubin. 2009. Variability and uncertainty in the cost of saline formation storage. *Energy Procedia* 1(1): 4151–4158.
93 Gresham, R. Lee. 2010. Chapter 4 in *Geologic CO_2 Sequestration and Subsurface Property Rights: A Legal and Economic Analysis*, doctoral thesis, Department of Engineering and Public Policy, Carnegie Mellon.
94 McCoy and Rubin, Variability and uncertainty.
95 Calabresi, G., and A. D. Melamed. 1975. Property rules, liability rules, and inalienability: One view from the cathedral. 85 *Harv. L. Rev.* 1089 (1975) (some entitlements are protected by a "property rule" (i.e., an injunction), which permits violation of the entitlement only with permission of the property owner).
96 See *Loretto v. Teleprompter Manhattan CATV Corp.*, 458 U.S. 419, 436–38 (1992); and *Tahoe-Sierra Pres. Council, Inc. v. Tahoe Reg'l Planning Agency*, 535 U.S. 302, 322 (2002).
97 For a discussion of approaches to determine value and just compensation, see Wilson and de Figueiredo, Geologic carbon dioxide sequestration; Klass and Wilson, Climate change, carbon sequestration, and property rights; Anderson, Subsurface "trespass"; and Anderson and Gresham, *Legal and Commercial Models*.
98 See S.B. 498, 61st Leg., Reg. Sess. (Mont. 2009) (codified at Mont. Code Ann. § 82–11–180 (2009)); S.B. 2319, 61st Leg. Assem., Reg. Sess. (N.D. 2009) (codified at N.D.

Cent. Code § 47–31–01–08 (2006 & Supp. 2009)); H.B. 89, 59th Leg., Budget Sess. (Wyo. 2008) (codified at Wyo. Stat. Ann. § 34–1–152 (2009)).

99 Interstate Oil and Gas Compact Commission (IOGCC). 2007. *Storage of Carbon Dioxide in Geologic Structures: A Legal and Regulatory Guide for States and Provinces* (hereinafter referred to as the IOGCC Model Statute).

100 See note 99.

101 *International Salt v. Geostow*, 697 F. Supp. 1258 (W.D.N.Y 1988); *Dept. Transp. v. Goike*, 560 N.W.2d 365 (Mich. Ct. App. 1996).

102 See note 99.

103 US Constitution Amendment V.

104 *Loretto v. Teleprompter Manhattan CATV Corp.*, 458 U.S. 419 (1982).

105 US Constitution Amendment V.

106 *Brown v. Legal Found. of Wash.*, 538 U.S. 216 (2003).

107 See *Chance v. BP Chems. Inc.*, 670 N.E.2d 985 (Ohio 1996); as well as Wilson and de Figueiredo, Geologic carbon dioxide sequestration; Klass and Wilson, Climate change, carbon sequestration, and property rights; and Anderson and Gresham, *Legal and Commercial Models.*

108 *Costal Oil & Gas Corp. v. Garza Energy Trust*, 268 S.W. 3d 1 (Tex. 2008); *Chance v. BP Chems. Inc.*, 670 N.E.2d 985 (Ohio 1996).

109 See note 108.

110 Sprankling, J. G. 2008. Owning the center of the earth. 55 *UCLA L. Rev.* 979 (2008).

111 *Costal Oil & Gas Corp. v. Garza Energy Trust*, 268 S.W. 3d 1 (Tex. 2008); *Chance v. BP Chems. Inc.*, 670 N.E.2d 985 (Ohio 1996).

112 Ibid.

113 Ibid.

114 Ibid.

115 42 U.S.C. § 171.

116 *Getty Oil Co. v. Jones*, 470 S.W.2d 618 (Tex. 1971); *Sun Oil Co. v. Whitaker*, 483 S.W.2d 808 (Tex. 1972).

117 15 U.S.C. § 717f(h).

118 See Wilson and de Figueiredo, Geologic carbon dioxide sequestration; Klass and Wilson, Climate change, carbon sequestration, and property rights; Anderson, Subsurface "trespass"; and Anderson and Gresham, *Legal and Commercial Models.*

119 *Columbia Gas Transmission Corp. v. Exclusive Natural Gas Storage Easement*, 962 F.2d 1192 (6th Cir. 1992).

120 See note 118.

121 *Lucas v. S.C. Coastal Council*, 505 U.S. 1003 (1992).

122 Calabresi, G., and A. D. Melamed. 1975. Property rules, liability rules, and inalienability: One view from the cathedral. 85 *Harv. L. Rev.* 1089 (1975) (some entitlements are protected by a "liability rule" (i.e., damages) which permits violation of the entitlement without permission of the owner so long as the violator pays damages).

123 US Department of Energy (DOE). 2008. *Carbon Sequestration Atlas of the United States and Canada.*

124 42 U.S.C. §4321.

125 16 U.S.C. §470.

126 16 U.S.C. §1531 *et seq.*

127 US Department of the Interior (DOI). 2009. *Framework for Geological Carbon Sequestration on Public Land. Report to Congress in Compliance with Section 714 of the Energy Independence and Security Act of 2007* (P.L. 110–140, H.R.6).

128 43 U.S.C. §1701 *et seq.*

129 Ibid.

130 See 30 U.S.C. §181 *et seq*; and 30 U.S.C. §351 *et seq.*

131 US Department of the Interior, *Framework for Geological Carbon Sequestration.*

132 42 U.S.C. § 1761(a)(2), (4),(7).

133 Senate Report No. 94–583.
134 Ibid.
135 Ibid.
136 3 U.S.C. § 1732(b); 43 C.F.R. § 2920.1–1.
137 US Department of the Interior, *Framework for Geological Carbon Sequestration*.
138 30 U.S.C. § 226(m).
139 *Exxon Corp. v. Lujan*, 970 F.2d 757 (10th Cir. 1992).
140 Ibid.
141 *Exxon Corp. v. Lujan*, 730 F.Supp. 1535 (D.Wyo. 1990).
142 Ibid.
143 Ibid.
144 Ibid.
145 Ibid.
146 40 C.F.R §§ 144–146; 42 U.S.C. § 300(h)(d) (2006); US Environmental Protection Agency (EPA). 2010. *Federal Requirements Under the Underground Injection Control (UIC) Program for Carbon Dioxide (CO_2) Geologic Sequestration (GS) Wells: Final Rule*. Fed. Reg. 75: 77230–303. December 10.
147 US Environmental Protection Agency (EPA). 2010. *Underground Injection Control Program: UIC Program Primacy*. http://www.epa.gov/safewater/uic/primacy.html#who (accessed July 29, 2010).
148 The "subsurface project boundary" is the ex ante estimated special extent of free-phase injected CO_2 delineated by both the lateral and the vertical boundaries from the time injection commences to the time that free-phase CO_2 ceases flowing, and taking into account a margin of error in predictions.

7 LIABILITY AND THE MANAGEMENT OF LONG-TERM STEWARDSHIP

1 See, for example, World Resources Institute (WRI). 2006. *Summary of WRI Workshop on Carbon Capture and Storage Liability* available at: http://pdf.wri.org/css_liability_092906.pdf [hereinafter WRI CCS Liability Workshop] (noting that industry representatives would not undertake CCS projects without regulatory clarity and perhaps even a federal indemnity program).
2 Price-Anderson may not be the best model for liability issues arising from long-term stewardship of CO_2 because it is principally directed at risks that deal with conditions that occur during Operation, rather than liabilities arising after nuclear plant decommissioning (the analog to long-term stewardship). See Trabucchi, Chiara and Lindene Patton. 2008. Storing carbon: Options for liability risk management, financial responsibility. *World Climate Change Report*, September 3, 2008, at 18 (noting that Congress's attempt to create an express financial risk management system for long-term stewardship of nuclear waste is fraught with problems associated with deferring the specifics of long-term siting and cost management for spent nuclear fuel to some future, undefined date).
3 See Hovorka, S. D., J. W. Choi, et al. 2009. Comparing carbon sequestration in an oil reservoir to sequestration in a brine formation—Field study. *Energy Procedia* 1(1): 2051–2056 (noting the similarities between EOR and aquifer sequestration).
4 *The Future of Coal Under Climate Legislation: The Importance of Risk Management in the Commercial Deployment of CCS: Hearing Before the H. Subcomm. on Energy and Environment*, 111th Cong. 2–3 (2009) (statement of Lindene E. Patton, Chief Climate Product Officer, Zurich Financial Services) [hereinafter Patton Testimony] (explaining that Zurich's CCS Liability Insurance and Geologic Sequestration Financial Assurance policies would be able to underwrite certain quantifiable elements of a CCS project, but that this did not include the post-closure phase for which risk is harder to anticipate); see also Patton,

Lindene. 2009. CCS: Risk management and insurance. *Greenhouse Issues* (March 2009), at 7–9 (explaining two new CCS insurance policies that Zurich will offer for protection against some of the liabilities associated with CCS projects. The CCS Liability Insurance policy is a geologic reservoir specific policy comprising five coverage grants including: pollution event liability, business interruption, control of well, transmission liability, and geomechanical liability. The second policy, Geologic Sequestration Financial Assurance (GSFA) insurance, is a "claims made and reported" coverage that provides cover for costs of performance of tasks specified in closure or post-operational plans. The maximum limit on each policy is $50 million).

5 The specific recommended standard is as follows: The project operator shall remain liable to the Board for any compensation paid or remediation cost incurred as a result of (A) the project operator's willful failure to comply with any regulatory requirement; (B) the reliance by the federal or state agency making the stewardship determination on any untrue statement of a material fact by the project operator; or (C) the omission by the project operator of a statement of material fact necessary in order to make the statements made not misleading (in light of the circumstances under which they were made) to the federal or state agency.

6 Note that several concepts in this paragraph are drawn from Klass, A. B., and E. J. Wilson. Fall 2008. Climate change and carbon sequestration: Assessing a liability regime for the long-term storage of carbon dioxide. *Emory Law Review* 58(103).

7 See Abraham, K. S. 1988. Environmental liability and the limits of insurance. 88 *Colum. l. Rev.* 942, 949.

8 See Anderson, D. R. 1978. Limits on liability: The Price Anderson Act versus other laws. *Journal of Risk and Insurance* 45(4): 651–674; and Katzman, M. T. 1988. Pollution liability insurance and catastrophic environmental risk. *Journal of Risk and Insurance* 55(1): 75–100.

9 Garner, Bryan A., ed. 2009. *Black's Law Dictionary*. 9th edition. Eagan, MN: West Publishing Company.

10 See 40 C.F.R. §§ 144–146.

11 Workers' compensation laws will be discussed in multiple sections because they incorporate multiple forms of statutory modified liability.

12 42 U.S.C. § 2210 (2006).

13 42 U.S.C. § 2210(c).

14 42 U.S.C. § 2210(o).

15 Trans-Alaska Pipeline Authorization Act (TAPAA), Pub. L. No. 93–153, § 201, 87 Stat. 576, 586–88 (1973).

16 Trans-Alaska Pipeline Authorization Act (TAPAA), Pub. L. No. 93–153, § 204(a), (c)(1)–(3).

17 Trans-Alaska Pipeline Authorization Act (TAPAA), Pub. L. No. 93–153, § 204(a)(2), (c)(2). The Oil Pollution Act of 1990, which was enacted by Congress to replace TAPAA following the *Exxon Valdez* oil spill, also allows for the removal of damage caps when an incident is caused by gross negligence or willful misconduct. 33 U.S.C. § 2704(c).

18 33 U.S.C. § 2704(c).

19 King, R. O. 2010. *Deepwater Horizon Oil Spill Disaster: Risk, Recovery, and Insurance Implications*. Congressional Research Service.

20 Greenwire. 2010. *Gulf Spill: BP to Waive Cap on Spill-Related Economic Damage Claims*. http://www.eenews.net/Greenwire/2010/10/19/9.

21 Note that concepts in this paragraph are drawn from Klass and Wilson, Climate change and carbon sequestration, 49.

22 See, for example, Longshore and Harbor Workers' Compensation Act, 33 U.S.C. §§ 901–944 (providing fixed awards to employees or their dependents in case of employment-related injuries or deaths occurring on navigable waters).

23 National Childhood Vaccine Injury Act of 1986, 42 U.S.C. §§ 300a-1 to 300aa-34 (creating no-fault compensation program for childhood vaccine-injury victims funded

by an excise tax on each dose of vaccine). See also: Rabin, R. 2005. The renaissance of accident law plans revisited. 64 *MD. L. Rev.* 698, 706–07 (discussing federal childhood-vaccine injury program).

24 See DOJ Vaccine Compensation Program—About the Program, at http://www.justice.gov/civil/common/vicp.html and HRSA-National Vaccine Injury Compensation Program, found at http://www.hrsa.gov/vaccinecompensation.

25 See DOJ Vaccine Compensation Program—About the Program, at http://www.justice.gov/civil/torts/const/vicp/about.htm.

26 The Oil Spill Liability Trust Fund, National Pollution Funds Center, found at http://www.uscg.mil/ccs/npfc/About_NPFC/osltf.asp.

27 Berkovitz, D. M. 1989. Price-Anderson Act: Model compensation legislation?—The sixty-three million dollar question. 13 *Harv. Envtl. L. Rev* 1, 2.

28 Atomic Energy Act, § 170(c); 42 U.S.C. § 2210(c). Note that concepts in this paragraph are drawn from Klass and Wilson, Climate change and carbon sequestration, 48ff.

29 Deutch, J., et al. 2003. *The Future of Nuclear Power: An Interdisciplinary MIT Study*. Massachusetts Institute of Technology. 81–83.

30 Berkovitz, Price-Anderson Act.

31 See generally 27 *Am. Jur. Indemnity* §§ 1–3 (1940).

8 GREENHOUSE GAS ACCOUNTING FOR CCS

1 See, for example, Kartha, S., M. Lazarus, et al. 2004. Baseline recommendations for greenhouse gas mitigation projects in the electric power sector. *Energy Policy* 32(4): 545–566; and Steenhof, P. 2009. Estimating greenhouse gas emission reduction potential: An assessment of models and methods for the power sector. *Environmental Modeling and Assessment* 14(1): 17–28.

2 We adopt the terminology "equipment leakage" and "surface leakage" in this chapter because these are the terms used by the rules governing GHG emissions reporting from GS sites.

3 US EPA. 2010. *Prevention of Significant Deterioration and Title V Greenhouse Gas Tailoring Rule*. In 40 C.F.R. aparts 51, 52, 70, and 71, Fed. Reg. Vol 75, No. 106, 2010; pp. 31514–31608.

4 US EPA. 2010. *PSD and Title V Permitting Guidance for Greenhouse Gases*. Fed. Reg. Vol. 75, No. 221, Nov 17, 2010, pp. 70254–70256. http://frwebgate1.access.gpo.gov/cgi-bin/PDFgate.cgi?WAISdocID=ujX1OG/2/2/0&WAISaction=retrieve (accessed February 9, 2011).

5 US EPA. 2010. *Available and Emerging Technologies for Reducing Greenhouse Gas Emissions from Coal-fired Electric Generating Units*. http://www.epa.gov/nsr/ghgdocs/electricgeneration.pdf (accessed February 9, 2011).

6 US EPA. 2009. *Mandatory Greenhouse Gas Reporting: Final Rule*. Fed. Reg. Vol. 74 No. 209, Fri. Oct. 30, 2009; p. 56260.

7 Fed. Reg. Vol. 74, No. 209, Oct. 30, 2009, pp. 56374–56519.

8 US EPA. 2010. *Mandatory Reporting of Greenhouse Gases: Petroleum and Natural Gas Systems: Final Rule*. Fed. Reg. Vol 75, No. 229, Nov. 30, 2010.

9 US EPA. 2010. *Final Mandatory Reporting of Greenhouse Gases from Carbon Dioxide Injection and Geologic Sequestration*. In 40 C.F.R. Part 98.440, Dec. 1, 2010; and Fed. Reg. Vol. 75 No. 230, 2010; pp. 75060–75089.

10 US EPA. 2009. *Mandatory Greenhouse Gas Reporting: Final Rule*. Fed. Reg. Vol. 74 No. 209, Fri. Oct. 30, 2009, pp. 56260–56373.

11 EPA acknowledges that establishing a baseline is impossible at EOR sites, and proposes alternatives.

12 See, for example, Teng, F., and D. Tondeur. 2007. Efficiency of carbon carbon storage with leakage: Physical and economical approaches, *Energy*, 32(4), April 2007; H. Herzog,

K. Caldeira and J. Reilly. 2003. An issue of permanence: Assessing the effectiveness of temporary carbon storage, *Climatic Change*,Volume 59(Number 3): 293–310; Hepple, R., and S. M. Benson. 2002. Implications of surface seepage on the effectiveness of geologic storage of carbon dioxide as a climate change mitigation strategy. In *6th International Conference on Greenhouse Gas Control Technologies*. Gale, J., and Y. Kaya, eds. Kyoto, Japan: Elsevier Science Ltd., pp. 261–266; and Ha-Duong, M., and D. W. Keith. 2003. Carbon storage: the economic efficiency of storing CO_2 in leaky reservoirs. *Clean Technologies and Environmental Policy* 5: 181–189.

13 See note 12.

14 van der Zwaan, B., and R. Gerlagh. 2008. *The Economics of Geological CO_2 Storage and Leakage*. Milan, Italy: Fondazione Eni Enrico Mattei.

15 See Benson, S. M. 2006. Monitoring carbon dioxide sequestration in deep geological formations for inventory verification and carbon credits. In *2006 SPE Annual Technical Conference and Exhibition*. San Antonio, TX: Society of Petroleum Engineers; and WRI. 2008. *Guidelines for Carbon Dioxide Capture, Transport and Storage*; Washington, DC: World Resources Institute. http://pdf.wri.org/ccs_guidelines.pdf.

16 Upp, E. L., and P. J. LaNasa. 2002. *Fluid Flow Measurement: A Practical Guide to Accurate Flow Measurement*. 2nd edition. Boston: Gulf Professional Publication.

17 Fed. Reg. Vol 75 No 230 Dec. 1, 2010 p. 75085 section 98.448(a)(3).

18 Benson, Monitoring carbon dioxide sequestration.

19 Benson, S. M., E. Gasperikova, and M. Hoversten. 2004. Overview of monitoring techniques and protocols for geologic storage projects. *IEAGHG Report*. No. PH4/29.

20 See Benson, Monitoring carbon dioxide sequestration; Cortis, A., C. M. Oldenburg, and S. M. Benson. 2008. The role of optimality in characterizing CO_2 seepage from geologic carbon sequestration sites. *International Journal of Greenhouse Gas Control* 2(4): 640–652.

21 See Leuning, R., D. Etheridge, A. Luhar, and B. Dunse. 2008. Atmospheric monitoring and verification technologies for CO_2 geosequestration. *International Journal of Greenhouse Gas Control* 2(3): 401–414; and Oldenburg, C. M., and A. J. A. Unger. 2004. Coupled vadose zone and atmospheric surface-layer transport of carbon dioxide from geologic carbon sequestration sites. *Vadose Zone Journal* 3(3): 848–857.

22 Lewicki, J., J. Birkholzer, and C.-F. Tsang. 2007. Natural and industrial analogues for leakage of CO_2 from storage reservoirs: Identification of features, events, and processes and lessons learned. *Environmental Geology* 52(3): 457–467.

23 Barr, J., S. Humphries, A. Nehrir, K. Repasky, L. Dobeck, J. Carlsten, et al. 2011. Laser-based carbon dioxide monitoring instrument testing during a 30-day controlled underground carbon release field experiment. *International Journal of Greenhouse Gas Control*, 5: 138–145.

24 Benson, Monitoring carbon dioxide sequestration.

25 Cooper, C., I. Wright, P. Ringrose, et al. 2009. A Technical Basis for Carbon Dioxide Storage. http://www.co2captureproject.org/co2_storage_technical_book.html (accessed October 5, 2011); Benson, S. M., E. Gasperikova, and M. Hoversten. 2004. Overview of monitoring techniques and protocols for geologic storage projects. *IEAGHG Report*. No. PH4/29; FutureGen Alliance. 2006. *Mattoon Site Environmental Information Volume*.

26 An example of applying geophysical methods to estimate the mass of CO_2 in place can be found in Nooner, S. L., O. Eiken, C. Hermanrud, et al. 2007. Constraints on the in situ density of CO_2 within the Utsira formation from time-lapse seafloor gravity measurements. *International Journal of Greenhouse Gas Control* 1(2): 198–214. For broader discussion, see Benson, Monitoring carbon dioxide sequestration in deep geological formations.

27 See Benson, S. M., M. Hoversten, E. Gasperikova, and M. Haines. 2005. Monitoring protocols and life-cycle costs for geologic storage of carbon dioxide. In *Proceedings of 7th International Conference on Greenhouse Gas Control Technologies*, Volume II, eds. E. S. Rubin, D. W. Keith and C. F. Gilboy, 1259–1265. IEA Greenhouse Gas Programme,

Vancouver, BC; Holloway, S., J. M. Pearce, V. L. Hards, T. Ohsumi, and J. Gale. 2007. Natural emissions of CO_2 from the geosphere and their bearing on the geological storage of carbon dioxide. *Energy* 32(7): 1194–1201; IPCC. 2005. *IPCC Special Report on Carbon Dioxide Capture and Storage. Prepared by Working Group III of the Intergovernmental Panel on Climate Change*. Cambridge: Cambridge University Press; and Lewicki, J., G. Hilley, L. Dobeck, and L. Spangler. 2010. Dynamics of CO_2 fluxes and concentrations during a shallow subsurface CO_2 release. *Environmental Earth Sciences* 60(2): 285–297.

28 Benson et al., Underground geological storage.

29 US EPA. 2009. *Mandatory Greenhouse Gas Reporting*. Fed. Reg. Vol. 74 No. 209, Fri. Oct. 30, 2009; Vol. 40 C.F.R. part 98, pp. 56374–56519.

30 See Benson, S., P. Cook, J. Anderson, S. Bachu, H. B. Nimir, B. Basu, et al. 2005. Underground geological storage. In *IPCC Special Report on Carbon Dioxide Capture and Storage*. B. Metz, O. Davidson, H. de Coninck, M. Loos, and L. Meyer, eds. Cambridge: Cambridge University Press; and Ide, T. S., K. Jessen, and F. M. Orr, Jr. 2007. Storage of CO_2 in saline aquifers: Effects of gravity, viscous, and capillary forces on amount and timing of trapping. *International Journal of Greenhouse Gas Control* 1(4): 481–491.

31 As, for example, under the American Clean Energy and Security Act of 2009, H.R. 2454 (*Waxman-Markey*) and the draft Senate Clean Energy Jobs and American Power Act (*Kerry-Boxer*).

32 Discussion of offsets typically includes the caveat that GHG reductions result from an activity that would not have been done otherwise (i.e., additionality), that the activity does not shift emissions to some other source (e.g., see Lokey, E. 2009. Valuation of carbon capture and sequestration under greenhouse gas regulations: CCS as an offsetting activity. *The Electricity Journal* 22(7): 37–47), and that the accounting framework captures emissions associated with parasitic energy consumption of capture equipment. On the one hand, as long as the capture facility is under the cap, concerns over additionality and emissions from parasitic energy consumption at the capture facility do not have to be considered. If, on the other hand, the capture facility falls outside the cap (e.g., in an uncapped sector or from a facility with emissions below the minimum), additionality and parasitic energy consumption must be addressed if offsets are generated.

33 See, for example, Emission Performance Standard, S.B. 1368, 2005–2006 Reg. Sess. (ca. 2006) (limiting long-term investments in baseload generation by the state's utilities to power plants that meet an emissions performance standard (EPS) jointly established by the California Energy Commission and the California Public Utilities Commission), codified at *Cal. Pub. Util. Code* §§ 8340(h), 8341 (West 2008); *Order Instituting Rulemaking to Implement the Commission's Procurement Incentive Framework and to Examine the Integration of Greenhouse Gas Emissions Standards into Procurement Policies, Cal. Pub. Util. Comm.*, Rulemaking 06–04–009 (Jan. 25, 2007) (adopting an EPS as directed by SB 1368).

34 Rubin, E. S. 2009. *A Performance Standards Approach to Reducing CO_2 Emissions from Electric Power Plants*. Pew Center on Global Climate Change. http://www.pewclimate.org/docUploads/Coal-Initiative-Series-Rubin.pdf (accessed February 9, 2011).

35 As, for example, under the American Clean Energy and Security Act of 2009, H.R. 2454 (*Waxman-Markey*) and the draft Senate Clean Energy Jobs and American Power Act (*Kerry-Boxer*).

36 This is the approach taken by *Waxman-Markey* and *Kerry-Boxer* (all GS sites would be considered "covered entities" subject to emissions caps) and by the draft protocol for GS under the EU-ETS.

37 See Chapter 7.

9 MAKING CCS A REALITY

1 Alic, J. A., D. S. Mowery, and E.S. Rubin. 2003. *U.S. Technology and Innovation Policies: Lessons for Climate Change.* Washington, DC: Pew Center on Global Climate Change.
2 Patiño-Echeverri, D., B. Morel, J. Apt, and C. Chen. 2007. Should a coal-fired power plant be replaced or retrofitted? *Environmental Science & Technology* 41(23): 7980–7986.
3 26 U.S.C. §45(a).
4 H.R. 3090, The Job Creation and Worker Assistance Act of 2002, 107th Congress (2001–2002), became Public Law No: 107–147.
5 H.R. 6, The Energy Policy Act of 2005, 109th Congress (2005–2006), became Public Law No: 109–58.
6 H.R. 6111, Tax Relief and Health Care Act of 2006, 109th Congress (2005–2006), became Public Law: 109–432.
7 H.R. 4520, American Jobs Creation Act of 2004, 108th Congress (2003–2004), became Public Law No: 108–357.
8 26 U.S.C. §45(c)(7)(A)(i).
9 26 U.S.C. §45(c)(7)(A)(iii) & §45(c)(7)(B).
10 26 U.S.C. §45(c)(9).
11 Internal Revenue Service. Form 8835: Renewable Electricity Production Credit, p. 2.
12 Ibid.
13 U.S. Energy Information Administration Electric Power Annual 2006 (with data from 2005).
14 See http://www.fossil.energy.gov/sequestration/partnerships/index.html.
15 Indiana House Enrolled Act, Bill No. 1722.
16 Indiana House Enrolled Act, Bill No. 1722. "Substitute natural gas" means pipeline-quality gas produced by a facility that uses a gasification process to convert coal into a gas capable of being used by a utility to supply end-use consumers with gas utility service, or as a fuel used to produce electric power to supply electric utility service.
17 Indiana House Enrolled Act, Bill No. 1722.
18 Indiana House Enrolled Act, Bill No. 1722.
19 72 Pa. P.S. §8101(b)(1).
20 72 Pa. P.S. §8101(b)(2).
21 Pena, N., and E. S. Rubin. 2008. *A Trust Fund Approach to Accelerating Deployment of CCS: Options and Considerations.* Washington, DC: Pew Center on Global Climate Change. http://www.pewclimate.org/white_papers/coal_initiative/trust_fund; Rubin, E. S. 2007. Accelerating deployment of CCS at U.S. coal-fired power plants. In *Proceedings of the 6th Annual Conference on Carbon Capture and Sequestration.* Pittsburgh, PA.
22 See http://thomas.loc.gov/home/gpoxmlc110/h6258_ih.xml.
23 Lyon, T. P., and E.-H. Kim. 2006. *Greenhouse Gas Reductions or Greenwash?: The Doe's 1605b Program.* Working Paper, University of Michigan.
24 Taylor, M.R., E. S. Rubin, and D. A. Hounshell. 2005. Control of SO_2 emissions from power plants: A case of induced technological innovation in the U.S. *Technological Forecasting and Social Change* 72(6): 697–718.
25 Yeh, S., E. S. Rubin, M. R. Taylor, and D. A. Hounshell. 2005. Technology innovations and experience curves for nitrogen oxides control technologies. *Journal of the Air & Waste Management Association* 55(2): 1827–1838.
26 Rubin, E. S., S. Yeh, M. Antes, M. Berkenpas, and J. Davison. 2007. Use of experience curves to estimate the future cost of power plants with CO_2 capture. *International Journal of Greenhouse Gas Control* 1(2): 188–197.

INDEX